Magnetosphere and Solar Winds, Humans and Communication

Edited by Khalid S. Essa, Khaled H. Mahmoud and Yann-Henri Chemin

Published in London, United Kingdom

Magnetosphere and Solar Winds, Humans and Communication
http://dx.doi.org/10.5772/intechopen.91592
Edited by Khalid S. Essa, Khaled H. Mahmoud and Yann-Henri Chemin

Contributors
Wellen Rukundo, Oleg A. Troshichev, Peter Stauning, Natalya Kilifarska, Vladimir Bakhmutov, Galyna Melnyk, Vadim Bogdanov, Mikhail V. Shubov, Khalid S. Essa, Khaled H. Hussein Mahmoud, Taha M. Eldebss, Ouerfelli Ghofrane, Victor Christianto, Yunita Umniyati, Florentin Smarandache, Yann-Henri H. Chemin

First published in London, United Kingdom, 2022 by IntechOpen
IntechOpen is the global imprint of INTECHOPEN LIMITED, registered in England and Wales, registration number: 11086078, 5 Princes Gate Court, London, SW7 2QJ, United Kingdom

British Library Cataloguing-in-Publication Data
A catalogue record for this book is available from the British Library

Additional hard and PDF copies can be obtained from orders@intechopen.com

Magnetosphere and Solar Winds, Humans and Communication
Edited by Khalid S. Essa, Khaled H. Mahmoud and Yann-Henri Chemin
p. cm.
Print ISBN 978-1-80355-747-2
Online ISBN 978-1-80355-748-9
eBook (PDF) ISBN 978-1-80355-749-6

For EU product safety concerns:
IN TECH d.o.o., Prolaz Marije Krucifikse Kozulić 3, 51000 Rijeka, Croatia,
info@intechopen.com or visit our website at intechopen.com.

Meet the editors

Dr. Khalid S. Essa received his BSc with honors (1997), an MSc (2001), and a Ph.D. (2004) in Geophysics from the Faculty of Science, Cairo University, Egypt. He joined the staff of Cairo University in 1997 and was appointed a research professor of potential field methods in the Department of Geophysics in 2014. He completed post-doctoral visits at the University of Lorraine and Strasbourg University, France in 2018; Charles University, Prague in 2014; and Western Michigan University, USA in 2006. Dr. Essa is an editor and external reviewer for many top journals. He has attended several international geophysical conferences in the United States, Australia, and France. He is a member of the Society of Exploration Geophysicists (SEG), American Geophysical Union (AGU), American Association of Petroleum Geologists (AAPG), European Association of Geoscientists and Engineers (EAGE), and Egyptian Syndicate of Scientific Professions (ESSP). He is also a member of the Petroleum and Mineral Resources Research Council and the National Committee for Geodesy and Geophysics. In 2017, he received the Prof. Nasry Matari Shokry Award in Applied Geology from the Academy of Scientific Research and Technology and an award for Scientific Excellence in Interdisciplinary, Multidisciplinary, and Future Sciences from Cairo University.

Khaled Hussein Mahmoud, Ph.D., is a professor in the Department of Physics, Khurmah University College, Taif University, Saudi Arabia. He works in the field of polymers and polymer nanocomposites. His main research lies in studying the electrical, optical, and thermal properties of polymers. Dr. Khaled studied physics at Cairo University, Egypt. He also completed a professorship at the University of Cairo.

Dr. Yann-Henri Chemin, Ph.D. in remote sensing applications (AIT, Th, 2006), graduated in Planetary Sciences with Astronomy in 2016, after a full academic path in Earth Observation in Asia. He is now concentrating on security, defense, and space industries, a field filled with both commercial endeavors and highly technical advances. He worked on the hyperspectral analysis of the Apollo 12 landing site and its vicinity on the Moon's surface. He also worked on the thermodynamic energy balance of the surface-atmosphere interactions of the moon Titan and on the mapping of craters on the dwarf planet called Ceres. Dr. Chemin keeps a keen eye on the challenges to settle on the moon and Mars, especially food creation, settlement construction, and in situ resource utilization. His interests include philosophy, innovation, space science and exploration challenges, spatial sciences, and geopolitical impacts on space.

Contents

Preface

This book describes the magnetosphere and its effect on the solar wind, climate change, and magnetospheric disturbances (substorms and magnetic storms). It also examines solar wind sources and the sun and its physics of interactions within and at a distance on Earth with humans and communications.

This book includes ten chapters organized into two sections: "Magnetosphere" and "Solar Wind, Humans and Communication." Section 1 includes six chapters. Chapter 1 by Ghofrane et al. describes the magnetosphere and its effect on the solar wind. Chapter 2 by Kilifarska et al. provides synthesized information about geomagnetic field variability, particles' propagation in Earth's atmosphere, ion-molecular reactions initiating ozone formation in the lower stratosphere, as well as evidence for the validity of this chain of sequences, which gives an adequate explanation of hemispherical and longitudinal asymmetry of the lower stratospheric ozone distribution, regionality of climate change, and the formation of regional climate patterns, known as climatic modes. Chapter 3 by Stauning presents a systematic assessment of the correlation between various Polar Cap (PC) indices used in published works with the merging electric field in the solar wind and with ground-based global magnetic indices such as auroral indices and magnetospheric ring current indices. Chapter 4 by Troshichev demonstrates that the Polar Cap (PC) index responds to the changes in the electric field coupling with the magnetosphere. Moreover, it determines the development of magnetospheric disturbances such as substorms and magnetic storms. Chapter 5 by Rukundo discusses more recent developments and improvement challenges in the developed ionospheric models for predicting and forecasting space weather events by utilizing deep learning and neural network techniques. It also emphasizes explaining and understanding the energy transfer mechanisms from the solar wind to the Earth's atmosphere during storms, the effects of wind, and thermospheric composition. Chapter 6 by Bogdanov presents an application of the methods of nonequilibrium thermodynamics based on the combined principles of Prigogine and Onsager for the description of the linear approximation of transport processes in a collisionless plasma.

Section 2 includes four chapters. Chapter 7 by Chemin introduces the sun, its activity, and its material flux. Chapter 8 by Rukundo discusses solar proton activity over the solar cycle and its associated space radiation doses. It permits the reader to look deeper into the temporal variation of the sun's activity as perceived by instruments on Earth. Solar proton events and their measured doses have a direct impact on electronic equipment and the quality of communications on the Earth's surface and within the atmosphere. Chapter 9 by Shubov discusses the ramjet acceleration of microscopic black holes within stellar material, the thermodynamic exchanges during transit, and the velocity modifications from the matter of the sun. Chapter 10 by Christianto et al. examines macroscopic scales of spin, looking into how spin supercurrents could permit nonlocal interaction and synchronicity.

This book is a useful resource for scientists and researchers from various disciplines, especially those interested in studying the atmosphere, solar wind, and climatic changes.

Khalid S. Essa, Ph.D.
Professor,
Faculty of Science,
Geophysics Department,
Cairo University,
Egypt

Khaled H. Mahmoud, Ph.D.
Professor,
Physics Department,
Khurmah University College,
Taif University,
KSA

Yann-Henri Chemin, Ph.D.
Joint Research Centre,
European Commission,
Italy

Section 1

Magnetosphere

Chapter 1

Introductory Chapter: The Magnetosphere

Ouerfelli Ghofrane, Khaled H. Mahmoud, Taha M. Eldebss and Khalid S. Essa

1. Introduction

The magnetosphere is the region dominated by magnetic fields from several celestial objects. Which physical phenomena are organized. Any planet with a magnetic field Earth, Jupiter, Saturn, Uranus and Neptune has its own magnetosphere. The Earth's magnetosphere is located beyond the ionosphere that is to say above 800 to 1000 km altitude. If there were no solar wind, the Earth's magnetic spectrum would be similar to that of an isolated bar magnet. In reality, the magnetosphere acts as a screen and protects the earth's surface from excess solar wind, which is harmful to life. It opposes the solar wind like a bridge abutment diverts the current of a river. On the other hand, the solar wind distorts the magnetic spectrum of the Earth, giving it a comet shape.

In 1600 a study of earth's magnetosphere began, thanks to William gilbert who discovered that magnetosphere is a cavity globally shaped by the magnetic field of the Earth and modified by the solar wind. Outside the magnetosphere finds the magnetosheath, separated from it by the magnetopause. Amongst the solar wind and the magnetosheath, a shock wave is created and this forms a boundary called the bow shock. At the lower limit of the magnetosphere is the ionosphere. Many sub-regions make up the magnetosphere, such as the polar horns, the auroral zones, the plasmasphere, the plasma sheet.

In our solar system there are some planets that have a magnetosphere, which is the area round a planet controlled by the planets magnetic field. And earth has the strongest one of all. This field is the secret of the development of life on earth since it ensures the protection and the continuity of the ecological system.

The reason of the field can be elucidated by dynamo theory. He realized that earth s magnetic field is nearly a magnetic dipole, with the magnetic field pole near the earths geographic north pole and the other magnetic field near the earth geographic south pole. This lets the compass usable for navigation.

Amongst these regions, the plasmasphere is toroidal in shape and populated by cold plasma of mainly ionospheric origin: electrons and positive ions (90% H^+ protons, less than 10% helium He^+, oxygen O^+, as well as some traces of heavier ions). It extends on average to equatorial distances of about 4-5 RE. However, this radial extension varies according to the geomagnetic activity. Indeed, in the case of a sharp increase in this activity, its outer boundary, the plasmapause, can be 2 RE, or on the contrary extend beyond 7-8 RE during long periods of geomagnetic calm. These variations in activity also disturb the structure on a small and medium scale. Thus, plasmasphere plumes connected to the plasmasphere can shape and turn with it. Furthermore, the plasmapause may be dotted with structures of density at small scales, hence the new name describing this region: the boundary layer of plasmasphere.

The internal magnetic field of planet earth is so strong. If we omit disturbances, the earth's magnetic field of internal origin can be pictured in first approximation as a dipole having an axis titled by about 12° with respect to the axis of rotation (**Figure 1**). Besides, the magnetic pole near earth's geographic North Pole is essentially the south magnetic pole. When it comes to magnets opposites attract. The magnetic field points in the direction of the Earth's surface in the northern hemisphere, and to space in the southern hemisphere. Its magnitude at the surface of the Earth is of the order of 50 μT at the equator. The interplanetary medium is permanently swept by a wind of particles electrically charged from the Sun. This wind, baptized solar wind by [2], is very little dense (5 cm^{-3}), but very fast (400 to 800 km/s).

The geographic north and south poles mention the points where the earth's rotation axis intercepts earth's surface. If you are holding a tennis ball amongst your thumb and forefinger and push on the side to create it spin. The solar wind is now to be a mixture of materials initiate in the solar plasma, constructed of ionized hydrogen (electrons and protons) with an 8% component of helium and trace amounts of heavy ions.

The solar wind has been distinguished inner toward the sun to the orbit of Mercury, and outward past the orbits of Uranus and Neptune. The flux of particles The stream of charged particles from the upper atmosphere of the Sun (Corona). Compose of ~92% ionized hydrogen (electrons and protons), ~8% components of alpha particles (helium), and trace amounts of heavy ions and atomic nuclei, with energies in the range between 1.5 and 10 keV. Acceleration mechanism is still not fully clear thermal energy alone cannot account for the high speeds of the solar wind.

The interplanetary medium is permanently swept by a wind of particles electrically charged from the Sun. The solar wind alters the Earth's magnetic field, creating a cavity called the magnetosphere (so named by Gold [1959]). Under the influence of the solar wind and the magnetic field it transports (the interplanetary magnetic field, or IMF1), the magnetosphere is compressed on the side of the Sun, called day side, but very extended in the anti-solar direction, called side night (**Figure 2**). As a result, the

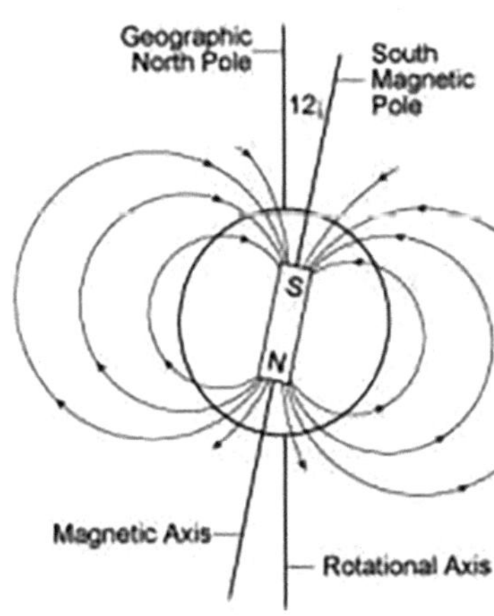

Figure 1.
Magnetic field of the earth in the absence of external disturbances [1].

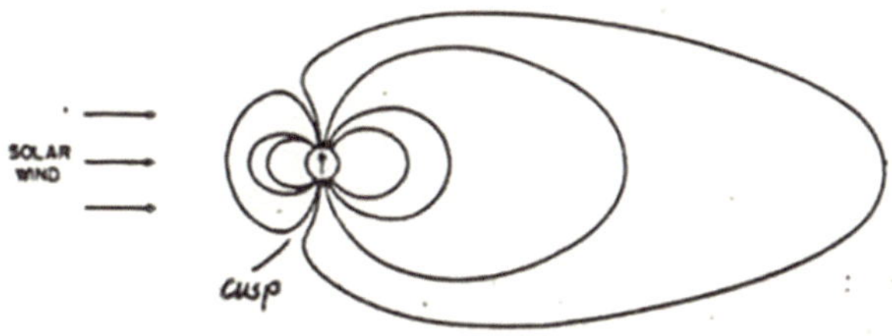

Figure 2.
The solar wind interaction with the Earth's magnetosphere prof. R. H. Holzworth, http://earthweb.ess.washington.edu/bobholz/ess515/.

distance amongst the outer boundary of the magnetosphere and the Earth is about 10 Earth radii (1 RE = 6371 km) from the dayside, while a tail magnetic field extending over several hundred terrestrial radii is formed on the night side [Ness, 1965].

2. The magnetosheath and the magnetopause

The magnetopause is the boundary between the magnetosphere, dominated by the magnetic field of the Earth, and the interplanetary medium, dominated by the solar wind (see the review by [3]). Its existence was first presented by [4]. The magnetopause is a relatively tight boundary in that it prevents the major part of the solar wind from entering the environment of the planet. The magnetosheath is a region in the vicinity of the magnetosphere, located amongst the shock and the magnetopause (see **Figure 3**). This is where the solar wind flows, mainly, bypassing the magnetosphere, after being slowed and heated through the impact. The plasma is denser there than in the wind solar upstream of the shock. It is a region where the plasma is turbulent, we measure a large electromagnetic stirring.

The magnetopause plays an important role in spatial physics, since the coupling amongst the solar wind and the magnetosphere is done through it. Note that the magnetosphere is not a static structure: it is in constant motion. On the one hand,

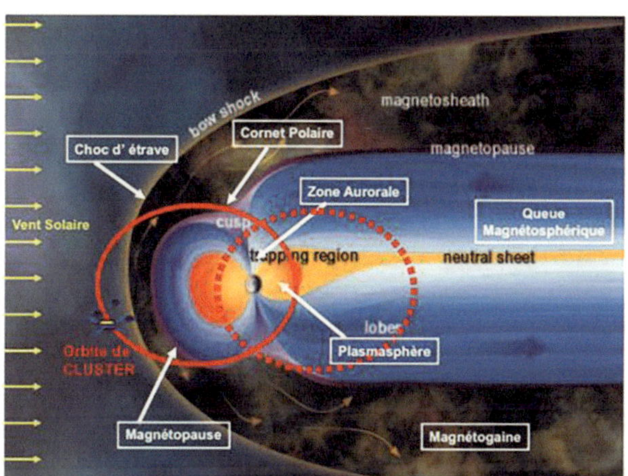

Figure 3.
Magnetosphere of the earth, with the main regions that compose it (according to ESA).

the orientation of Earth's magnetic dipole changes with its daily rotation and with its annual revolution around the Sun. On the other hand, the solar wind is characterized by a strong temporal variability on timescales ranging from seconds to years. The dimensions and shapes of these regions may change over time due to the natural variability.

The magnetopause is the boundary between the magnetosphere, dominated by the magnetic field of the Earth, and the interplanetary medium, dominated by the solar wind (see the review by Russell [3]). Its existence was first introduced by [4]. The magnetopause is a relatively tight boundary in that it prevents the major part of the solar wind from entering the environment of the planet. The magnetosheath is a region in the vicinity of the magnetosphere, located between the shock and the magnetopause (see **Figure 3**). This is where the solar wind flows, mainly, by passing the magnetosphere, after being slowed and heated through the impact. The plasma is denser there than in the wind solar upstream of the shock. It is a region where the plasma is turbulent; we measure a large electromagnetic stirring. Furthermore, the magnetopause plays an important role in spatial physics, since the coupling amongst the solar wind and the magnetosphere is done through it. Note that the magnetosphere is not a static structure: it is in constant motion. On the one hand, the orientation of Earth's magnetic dipole changes with its daily rotation and with its re- annual evolution around the Sun. On the other hand, the solar wind is characterized by a strong temporal variability on timescales ranging from seconds to years. The dimensions and shapes of these regions may change over time due to the natural variability. For example, when material from a solar corona eruption (a coronal mass ejection, or CME2) propagates through interplanetary space and reaches the Earth, the dynamic pressure of the solar wind is highly increased, so that the magnetopause is pushed toward the Earth. One of the effects of continuous minor fluctuations of the dynamic pressure of the solar wind is the oscillation movement of the magnetopause: this one oscillates in order to restore the balance between the magnetospheric pressure and the pressure of the solar wind. The movement of this boundary can also be induced by instabilities plasma. Spatio-temporal fluctuations are also known to make the magnetopause semi-permeable, and allow plasma of the magnetosheath to cross it, so forming the magnetospheric boundary layer. Various mechanisms allow this transfer of matter, such as magnetic reconnection [5], or the impulsive penetration of plasma elements [6].

2.1 Auroral zone

The auroral zones constitute two circular regions around the North magnetic poles and South, where auroras are commonly observed (**Figure 4**). Their geographical position was determined for the first time by [7], from a study statistics of aurora observations in the northern hemisphere. These light emissions re- result from the luminescence of the upper atmosphere because of the de-excitation of the molecules of the atmosphere. The molecules are previously excited by energetic electrons coming from the magnetosphere and precipitated into the atmosphere This region is also of immense interest because of the various types of waves that we observe, in the range of ultra-low frequencies, or ULF3 [Perraut et al., 1998], but also in the very low frequency range, or VLF4. They are mainly electrostatic waves, but some studies have also shown the existence of electromagnetic waves [8].

These waves play a huge important role in the areas auroral, and can be used to characterize the level of activity in this region of the magnetosphere. The results are

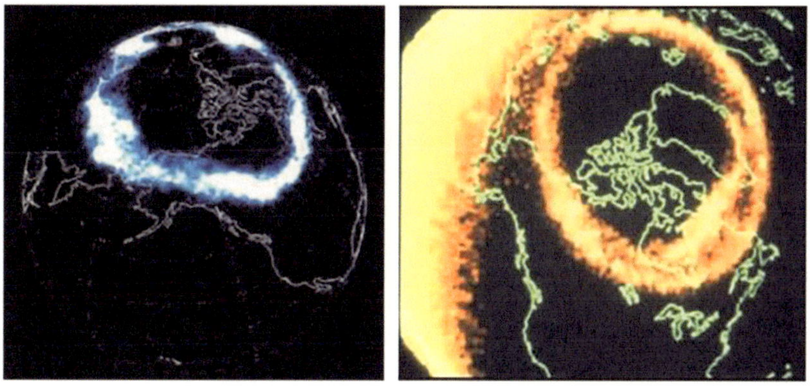

Figure 4.
Auroral oval in the northern hemisphere, observed by the IMAGE satellite (panel left), and by the polar satellite (right panel) (from NASA).

that, at mid- latitudes, VLF waves are known to disperse trapped energetic electrons, and eventually precipitate them into the Earth's atmosphere, where they can alter the propagation of radio waves.

2.2 The plasmasphere and the plasmapause

The internal magnetosphere notably includes the plasmasphere (**Figure 5**), which is a toroidal region surrounding the Earth, inhabited by cold plasma (a few eV or less) of mainly ionospheric origin (see the monograph by [9] and the review by [10]). In comparison with other regions of the magnetosphere, the density of the plasmasphere is really high (from 10 to 104 cm^{-3}). It is composed of electrons and positive ions: protons H+ (~ 90%), but also helium He^{+}, oxygen O^{+} and some traces of heavier ions.

This plasma is mainly of ionospheric origin, and during periods of low geomagnetic activity, a process of filling from the upper layers of the ionosphere let the

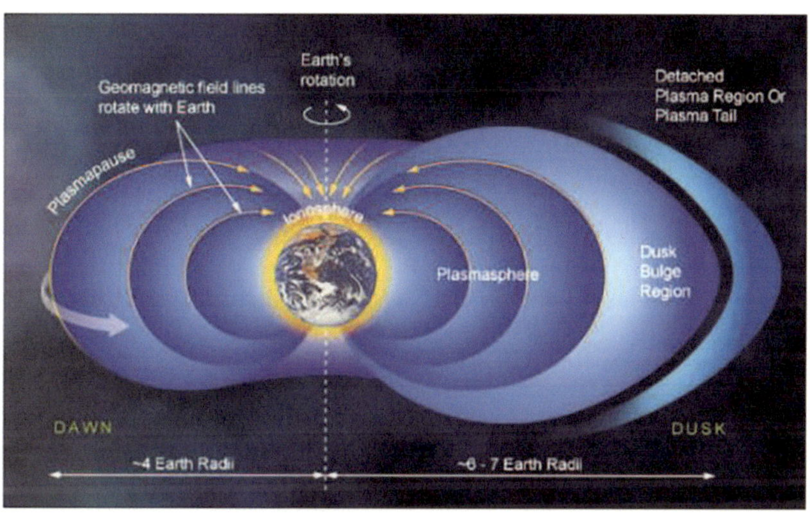

Figure 5.
General view of the plasmasphere (from "Windows to the Universe", http://www.windows.ucar.edu).

plasma to rise along the field lines magnetic and thus to fill the plasmasphere. This mechanism has not yet been fully elucidated and is currently the subject of numerous studies. The outer boundary of the plasmasphere was discovered simultaneously from ground observations of whistles [11], and from measurements on board satellites [12]; it was called plasmapause for the first time by D. Carpenter in 1966 [13]. It is characterized, in its simplest form, by a sudden drop in the plasma density by several orders of magnitude. Shape approaching that of a shell magnetic, this boundary can be located at the equator at a radial distance between 2 and 8 RE relying on the intensity of the geomagnetic activity. The plasmasphere occupies a region all the more extensive as the geomagnetic activity is weak. Plasmapause is not always a clear border, the density drop can be very irregular and occur on a wide range of geocentric distances. Carpenter and Lemaire [14] introduced recently the notion of "plasmasphere boundary layer", or PBL5, thick border which covers all the regions where the density drop occurs, roughly between 103 and 10 cm^{-3}. Finally, the plasmasphere has an extension on the twilight side that is to say in the local time.

Author details

Ouerfelli Ghofrane[1], Khaled H. Mahmoud[1], Taha M. Eldebss[2] and Khalid S. Essa[3]*

1 Department of Physics, Taif University, College of Khurma University College, Taif, Saudi Arabia

2 Faculty of Science, Chemistry Department, Cairo University, Giza, Egypt

3 Faculty of Science, Geophysics Department, Cairo University, Giza, Egypt

*Address all correspondence to: khalid_sa_essa@cu.edu.eg

References

[1] Lang KR. The Cambridge Guide to the Solar System. New York: Cambridge University Press; 2003. p. 468

[2] Parker EN. Dynamics of the interplanetary gas and magnetic fields. Astrophysical Journal. 1958;**128**:664

[3] Russell T. The structure of the magnetopause. Planetary and Space Science. 2003;**51**:731-744

[4] Chapman S, Ferraro VCA. A new theory of magnetic storms. Terrestrial Magnetism and Atmospheric Electricity. 1931;**36**:171-186

[5] Dungey JW. Interplanetary magnetic field and the auroral zones. Physical Review Letters. 1961;**6**:47-48

[6] Lemaire JF. Impulsive penetration of filamentary plasma elements into the magnetospheres of the earth and Jupiter. Planetary and Space Science. 1977;**25**:887-890

[7] Loomis E. The great auroral exhibition of August 28 to September 4, 1859—2nd article. The American Journal of Science. 1860;**29**:92-97

[8] Gustafsson G, Andre M, Matson L, Koskinen H. On wavesbelowthe localprotongyrofrequencyin the auroralaccelerationregion. Journal of Geophysical Research. 1990;**95**:5889

[9] Lemaire JF, Gringauz KI. The Earth's Plasmasphere. New York: Cambridge University Press; 1998. p. 372

[10] Ganguli G, Reynolds MA, Liemohn MW. The plasmasphere and advances in plasmaspheric research. Journal of Atmospheric and Solar-Terrestrial Physics. 2000;**62**:1647-1657

[11] Carpenter DL. Whistler evidence of a "knee" in the magnetospheric ionization density profile. Journal of Geophysical Research. 1963;**68**:1675-1682

[12] Gringauz KI. The structure of the ionized gas envelope of the earth from direct measurements in the USSR of local charged particle concentrations. Planet. Planetary and Space Science. 1963;**11**:281-296

[13] Carpenter DL. Whistler studies of the plasmapause in the magnetosphere: 1. Temporal variations in the position of the knee and some evidence on plasma motions near the knee. Journal of Geophysical Research. 1966;**71**:693-709

[14] Carpenter DL, Lemaire J. The plasmasphere boundary layer. Annals of Geophysics. 2004;**22**:4291-4298

Chapter 2

Coupling between Geomagnetic Field and Earth's Climate System

Natalya Kilifarska, Volodymyr Bakhmutov and Galyna Melnyk

Abstract

The idea about synchronized variations of geomagnetic field and climate appears in the middle of the twentieth century. Among others, one of the main reasons for its unpopularity is the missing mechanism of coupling between magnetic and non-magnetic media. This chapter offers such a mechanism, consisting of a chain of relations transmitting the geomagnetic spatial-temporal variations down to the planetary surface. The first element of this chain is energetic particles propagating in Earth's atmosphere, whose density and depth of penetration are modulated by geomagnetic field. Thus, the non-dipolar geomagnetic irregularities are projected on the ionization layer in the lower atmosphere (known as Regener-Pfotzer maximum). This unevenly distributed ionization, in certain conditions (i.e. dry atmosphere), acts as a secondary source of ozone near the tropopause. Ozone at this level is of special importance due to its influence on the tropopause temperature and humidity, and consequently on the planetary radiation balance. Hence, the geomagnetic spatial and temporal variations are imprinted down to the surface, impacting the climate system and its regional structures. The chapter provides synthesized information about geomagnetic field variability, particles' propagation in Earth's atmosphere, ion-molecular reactions initiating ozone formation in the lower stratosphere, as well as evidence for its covariance with some atmospheric variables.

Keywords: geomagnetic variations, geomagnetic focusing of charged particles, lower stratospheric ozone, regionality of climate changes

1. Introduction

The co-variability of paleomagnetic and paleoclimate time series has been found in many sedimentary records, e.g. [1] and references therein. Most of the reversals of geomagnetic field polarity and magnetic poles' excursions seem to appear in periods of cold climate [1, 2]. Other authors, however, announced that climatic cooling fairly well corresponds to episodes with a stronger geomagnetic field [3–5]. This controversy, together with objective difficulties for disentangling paleomagnetic from paleoclimate data – due to the high variability and climate dependence of marine sedimentation rates – determines the skepticism of the greater part of the scientific community regarding possible links between geomagnetic filed and climate.

On the other hand, time series based on contemporary instrumental measurements do not contain the ambiguity of paleo-data records. Based on the magnetic and climate measurements collected from the beginning of 1900 up to 2010, this chapter provides not only more evidence for existing coupling between geomagnetic field and climate system, but also offers a physically rational explanation and results supporting its validity.

2. Spatial-temporal variability of geomagnetic field at different time scales

The Earth's magnetic field interacts with all planetary shells – the core, mantle, and crust of the solid Earth, as well as with the atmosphere, hydrosphere, and biosphere. It comprises information about both the state of near-earth space and the internal structure of our planet. The Earth's magnetic field is continuously changing in space and time. The sources of its variations are located inside and outside of the planet. The amplitude and periodicity of geomagnetic variations are very different, which affects the methods used for data acquisition.

2.1 Long-term variations related to the heterogeneity at the core-mantle boundary

The longest periods of reoccurrence have geomagnetic *reversals*, followed by geomagnetic *excursions*. Geomagnetic *reversals* define the exchange of positions of the North and South magnetic poles. For the last million years, the geomagnetic field has changed its polarity four times. The last one happened about 780,000 years ago. During the inversion of geomagnetic polarity, the magnetic field's strength drops dramatically, leading to a severe weakening of the planetary magnetic shielding, which protects living organisms from harmful cosmic radiation. This is the argument of some scientists to suggest that episodes of mass extinctions of terrestrial biota could be attributed to geomagnetic reversal [6, 7]. No systematic pattern was found in the occurrence of inversions and they are treated as a random process.

The palaeomagnetic records reveals also the existence of shorter periods (with a duration of several thousand years) when the field has departed from its near-axial configuration. Such short-term events are called geomagnetic *excursions*. The *excursions* are usually defined as a deviation of the virtual geomagnetic pole equatorward of 45° latitude, or as a short-term change in the direction of a geomagnetic field, whose amplitude is at least three times greater than the *secular variations* for a given period of time. *Excursions* are short-term impulse fluctuations, which are mostly replaced by smoother secular variations in geomagnetic field intensity. The nature of geomagnetic polarity reversals and excursions is not fully understood. Their general characteristics suggest that they could be considered manifestations of various processes within the Earth's liquid core.

Secular variations are another long-term variability of geomagnetic field ranging from decades to several thousands of years. It is generally accepted that the geomagnetic secular variations are associated with changes at the core-mantle boundary. They are studied using all available methods – paleomagnetic, archeomagnetic, and direct observations. Over the period of instrumental observations (approximately 120 years), secular variations are grouped in 4 intervals: 60–70, 30–40, 18–25, 10–11 years. Variations with a period of about 60 years have the greatest amplitude.

In the first approximation, the magnetic field is interpolated as the field of a magnetic dipole. However, the empirical models (incorporating all available measurements of field intensity), as well as satellite measurements, reveal the existence of a non-dipolar component in the real geomagnetic field. The irregularities in the spatial distribution of geomagnetic field intensity are well visible in **Figure 1**, based on the 13th generation of the International Geomagnetic References Field model (IGRF) [8]. The two-wave distribution of field intensity in the Northern Hemisphere and a single-wave in the Southern Hemisphere are well visible in **Figure 1**.

Moreover, the temporal evolution of geomagnetic field also differs in different regions over the world. The greatest amplitude of changes is observed in the Western Hemisphere, in the regions of the Canadian (**Figure 2a**) and South Atlantic (**Figure 2c**) world anomalies. In the Eastern Hemisphere amplitudes of these changes are smaller (**Figure 2b** and **d**). The spatial structure of these irregularities is well visible in the maps of geomagnetic secular variations (**Figure 3**), which are calculated by the formula: $F_{sv} = (F_{t_2} - F_{t_1})/(t_2 - t_1)$, where F_{t_1} and F_{t_2} are field intensity in two moments in time, and $(t_2 - t_1)$ is the length of the period in years.

Figure 3 illustrates fairly well that focuses of the strongest secular variations evolve with time, in their strength and position over the globe. All the features of the spatio-temporal structure of the geomagnetic field, the problems of its observations and modeling, are described in great details in [9].

2.2 Heterogeneous interplanetary environment and its imprint on the geomagnetic field's short-term variability

Short-term changes in geomagnetic field (from seconds to days) are caused by the external sources – i.e. the current systems in the magnetosphere and ionosphere. In the absence of solar-terrestrial disturbances, the Earth's magnetic field shows regular daily variations with small amplitude (~tens of nT), which are primarily composed of 24, 12, 8, and 6-hour spectral components [10–12]. These variations are known as

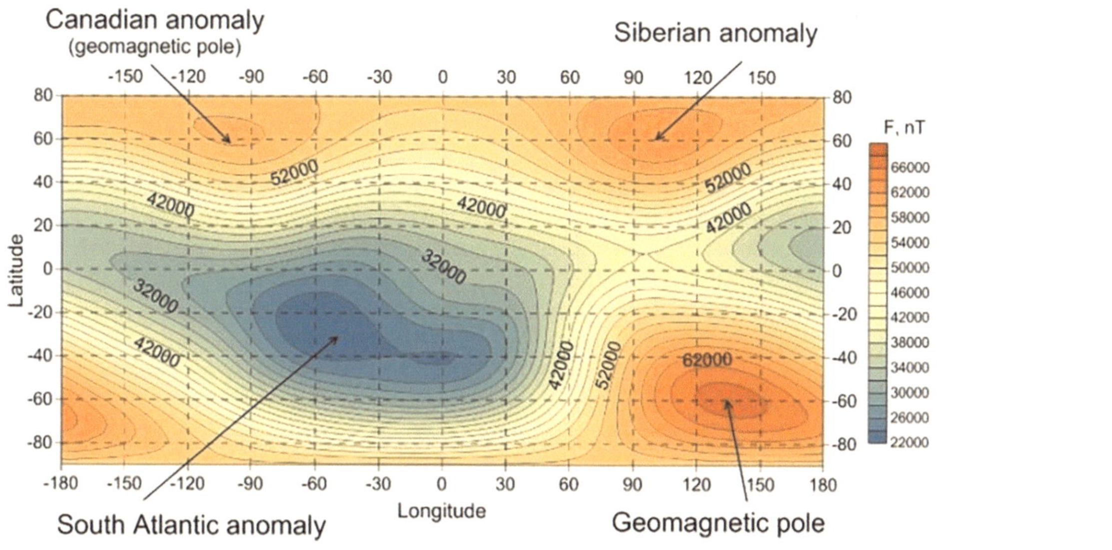

Figure 1.
Spatial structure of the modulus of the total vector of the geomagnetic field intensity, calculated for 2021 by the IGRF-13 model. (https://www.ngdc.noaa.gov/geomag/calculators/magcalc.shtml#igrfgrid).

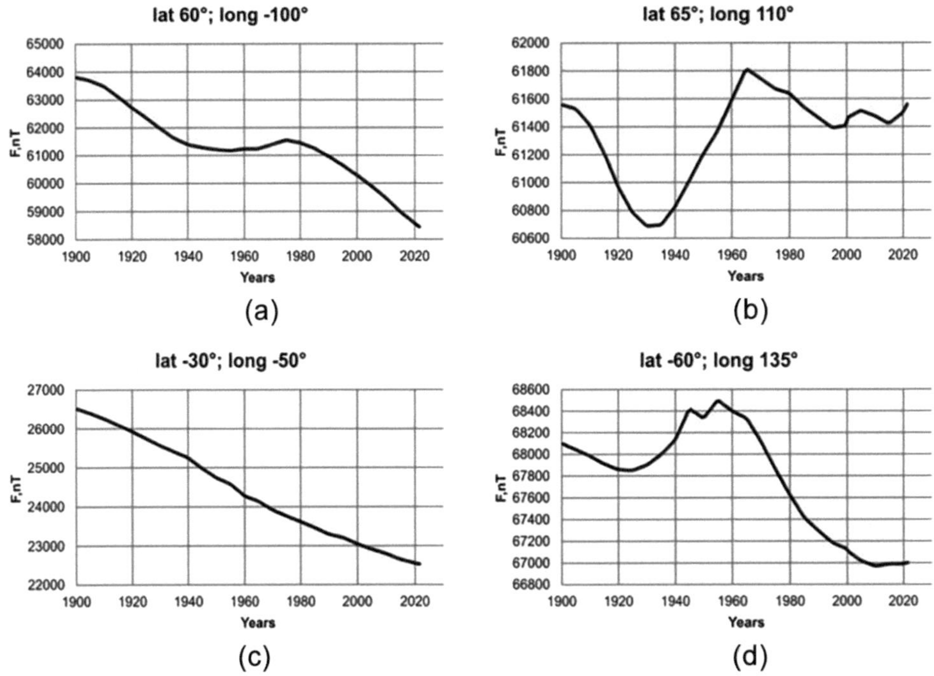

Figure 2.
Geomagnetic field changes in the regions of world geomagnetic anomalies: (a) Canadian, (b) Siberian, (c) South Atlantic and (d) Geomagnetic pole in the southern hemisphere.

solar quiet (Sq) variations. Today it is well understood that Sq variations are induced by the electric currents existing in the ionospheric dynamo region (between 90 and 150 km), where the neutral wind drives an electromotive force – through the ionospheric wind dynamo mechanism [13, 14]. The Sq variations are sensitive to the sunspot numbers [10]. For example, the midlatitude Sq currents' intensity is approximately twice higher in solar maximum than in solar minimum conditions [15, 16].

In addition, the geomagnetic perturbations at the planetary surface also have *lunar* spectral components. The stronger one is the semidiurnal lunar variation with a period of 12 hours in lunar time or ~12.42 hours in solar time. The typical amplitude of *lunar* variation is much smaller – approximately one-tenth of the Sq variation [17]. Geomagnetic lunar variability is a consequence of atmospheric lunar tides, inducing ionospheric currents in the ionospheric dynamo region, which are furthermore projected on the ground [18, 19].

The maximum amplitude of quiet Sq and lunar variations has a maximum during the daytime hours, and when the moon is in opposition. These are smooth periodic variations with intensities reaching 200nT, increasing from the equator to the poles [20].

The quiet conditions, however, are frequently disturbed by active processes on the Sun (e.g. solar flares, coronal mass ejection, coronal holes, etc.). The ejected solar mass and magnetic fields propagate in the interplanetary magnetic field (IMF) as a shock wave, which distorts significantly geomagnetic field when it splashes on the Earth. Only ~1% of energy carried by the solar wind is transferred to the Earth's magnetic field because the reconnection between interplanetary and geomagnetic fields depends on their directions. It is well established that the southward direction of the interplanetary magnetic field favors its reconnection with the Earth's magnetic field.

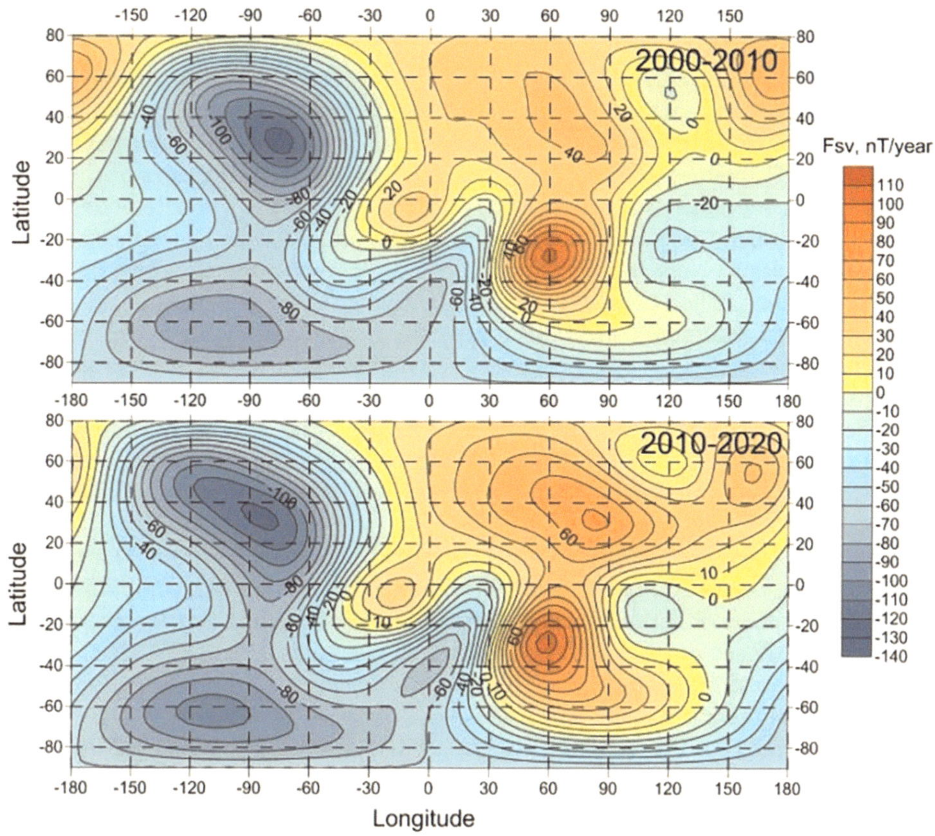

Figure 3.
Secular variations of the first two decades of twenty-first century, based on the IGRF-13 model. (https://www.ngdc.noaa.gov/geomag/calculators/magcalc.shtml#igrfgrid).

The energy transferred to the magnetosphere in such periods abrupt dramatically by one-two orders of magnitude, reaching power of ≥1011 W [21]. These periods are known as *geomagnetic storms* or *substorms.*

The dominant interplanetary phenomena causing *intense* magnetic storms (with an equatorial Dst index lower than −100 nT) depends on the solar cycle. Around the solar maximum, the interplanetary medium is dominated by fast coronal mass ejections (CMEs). Two interplanetary structures are important for the development of storms, involving intense southward IMFs: the *sheath* region just behind the forward shock, and the CME ejecta itself. Whereas the *initial phase* of a storm (manifesting itself as a sudden impulse in geomagnetic field) is caused by the increase of plasma pressure at, and behind the shock, the storm's *main phase* is due to southward IMFs. The storm recovery begins when the IMF turns less southward, with delays of ≈ 1–2 hours, and has typically a decay time of 10 hours [22].

Magnetic clouds are large-scale interplanetary formations, caused by coronal mass ejection on the Sun, in which the magnetic field strength, propagating speed, and plasma concentration are higher than in the surrounding flows [23]. The vertical Bz component of IMF slowly changes from negative to positive sign in SN clouds, and vice versa in NS clouds. The interaction of the Earth's magnetosphere with magnetic clouds, as a rule, is accompanied by intense geomagnetic disturbances [24, 25].

According to some estimates, the geoeffectiveness of magnetic clouds to disturb Earth's magnetic field is 77% [25, 26].

During solar minimum, high-speed streams from coronal holes dominate the interplanetary medium activity. The high-density, low-speed streams (associated with the heliospheric current sheet plasma) impinging upon the Earth's magnetosphere cause positive Dst values in the initial phase of the storm. In the absence of shocks, sudden impulses are infrequent in periods of low solar activity. The interaction between fast stream (emanated from coronal holes) and the slow heliospheric current sheet plasma leads to the formation of a compression region with a high magnetic gradient, called *Corotating Interaction Region* (CIR). The main phase of magnetic storms generated by the CIR is typically weaker and highly irregular. The recovery of geomagnetic storms that happened in periods of inactive Sun is also quite different – lasting from many days to weeks. The southward magnetic field component of Alfvén waves, existing in the high-speed plasma stream, causes intermittent reconnection and substorm activity, as well as sporadic injections of plasma sheet energy into the outer portion of the ring current, prolonging its final decay to quiet day values [22].

For certain classes of magnetic storms, the interaction of CIR with the Earth's magnetosphere is more efficient than CME [27]. On the other hand, comparisons of the geoeffectiveness of various interplanetary structures, such as shock waves, magnetic clouds, IMF sectors boundaries, and CIR, showed that 33% of CIR are accompanied by moderate or intense storms. This means that every third phenomenon of the observed CIR at the Earth is geoeffective [28].

It is statistically confirmed that geoeffective disturbances can be caused by a whole spectrum of various phenomena on the Sun: *flares* (especially with the release of high-energy protons); the *sudden disappearance of filaments*, followed by a transition to a *coronal mass ejection*; *high-speed streams* of solar wind; Earth passage through the IMF sectors' boundaries, etc. However, the features of the magnetic storms are primarily determined by changes in IMF and solar wind parameters [29–32].

The influence of geomagnetic storms on the lower atmospheric variables is studied by many authors. The storm imprint on the near-surface pressure and temperature has been reported by [33–34], on circulation by [35–40], on total ozone density by [41], etc. The latter authors have compared geomagnetic storm manifestation in upper, middle, and lower atmosphere, emphasizing on differences in the atmospheric response to geomagnetic storms. Their main conclusions are summarized as follow: (i) unlike the prevailing latitudinal dependence of storm impact on the upper-middle atmosphere, the tropospheric effects manifest itself with a well pronounced *regionality*; (ii) the weak seasonal dependence of the storm effect in the upper middle atmosphere is altered by a *strong seasonal dependence* of detected tropospheric response; and (iii) the geomagnetic effect in the upper middle atmosphere are caused primarily by energetic particles, while the *origin of tropospheric effect* is still not well understood [41].

All these effects are due to the short-term geomagnetic disturbances, initiated by the external influence – i.e. solar variability and inhomogeneity of interplanetary medium. Although important, these fluctuations of Earth's magnetic field are short-lasting and their impact on the climate system is negligible. Oppositely, this publication is focused on the long-term variations of geomagnetic field on interdecadal and multidecadal time scales (initiated at the core-mantle boundary) and their relation to climate variability with its regional specifics.

3. Geomagnetic modulation of charged particles in Earth's atmosphere

3.1 Van Allen radiation belts

Important structures in Earth's magnetosphere are its radiation belts, which consist of relativistic electron and proton populations, trapped in the Earth's magnetic field. The Earth has two such belts and some others may be temporarily created. The *outer* radiation belt, occupying the space between 3 and 10 Earth's radii (R_E), consists mainly of electrons with energies 0.1–10 MeV (million electron volts). The outer belt's particles have both – solar (mainly helium ions) and atmospheric origin. The protons of the *outer* belt, however, possess much lower energies than those of the *inner* belt. The most energetic particles of the outer belt are electrons, achieving energies of several hundred MeV.

An electrons population is found also in the outer edge of the *inner* radiation belt at a distance 1.5÷3 R_E [42]. The outer and inner electrons' belts are separated by the *slot* region, where the interactions with the electromagnetic waves called "whistlers" are the main reason for the lower density of the electron population. The electrons' loss in the slot region is due to the pith angle scattering (related to the impact of whistlers) – facilitating their escape from the geomagnetic trap [43]. The inner belt electron population is periodically refreshed by the transport of electrons from the outer radiation belt [44]. Moreover, it has been recently recognized that the decay of thermal energy neutrons, produced by cosmic rays striking the upper atmosphere, contributes to energetic electrons in the inner belt and acts as the dominant source of energetic electrons at the inner edge of the inner belt [45].

The *inner* radiation belt, occupying the near Earth space between 0.2 and 2 R_E, is largely populated, however, by energetic protons with energies exceeding 30 MeV. According to the current understandings, these protons originate from: (i) the decay of neutrons – produced within the interaction between galactic cosmic rays and atmospheric atoms and molecules [46], and (ii) solar protons – injected into the interplanetary space during the solar flares and coronal mass ejections [47–49]. The solar energetic protons are the primary source of particles for the inner belt, which energy is beneath ~100 MeV [49] and sometimes produces a long-lived proton belt – distinct from the inner radiation belt [47, 50].

3.2 Particles' focusing in a heterogeneous magnetic field

Particles trapped within the geomagnetic field are urged by the Lorentz force (1) to move along the magnetic field lines on spiral trajectories (the result of a combined *circular* and a *field align* motions), continuously bouncing between the Northern and the Southern Hemispheres.

$$m\frac{dv}{dt} = q[E + (v \times B)]; v = \frac{dr}{dt} \tag{1}$$

where: B(r,t) is external magnetic field – function of the spatial dimensions and time, r and v are respectively particle's radius vector and velocity; "m" is particle's mass and "q" – its charge.

Besides the helical movement of particles along geomagnetic field lines, they also perform the additional movement in a direction perpendicular to the magnetic field lines – known as *magnetic drift*. This type of motion is determined by the non-uniformity of Earth's magnetic field in the direction perpendicular to B, and by the magnetic field curvature (2)

$$v_{drift} = \frac{m}{q \cdot B^2}\left(v_{\perp}^2 \frac{B \times \nabla B}{2B} + v_{II}^2 \frac{\rho \times B}{\rho^2}\right) \quad (2)$$

where B is the magnetic vector, ρ – the radius of the geomagnetic lines curvature, v_{II} and $v_{\perp}$ are projections of particle's velocities parallel and perpendicular to geomagnetic field line; q and m are respectively particle's charge and mass. The first term in the brackets corresponds to the magnetic gradient perpendicular to the field lines, while the second term – to their curvature.

Formula (2) shows also that particles' drift across the magnetic field lines depends on their charge q, and consequently leads to a charge separation, which in turn generates electric field E along the drift direction. The combined effect of E and B fields induces an $E \times B/B^2$ drift of particles, which displaces positive ions and negative electrons in the same direction – perpendicular simultaneously to B and to E. These charged particles are then "lost" in the ambient atmosphere, where they release their energy, producing showers of secondary particles.

In a dipolar geomagnetic field (with its cross-latitudinal magnetic gradient) the protons are drifting westward, while electrons – eastward. The real geomagnetic field has, however, a non-dipole component creating additionally a cross-longitudinal gradient. In this case, the protons (entering the denser atmosphere from the west) are shifted sought-westward in regions with a positive cross-longitudinal gradient and sought-eastward – in regions with a negative gradient (refer to Eq. (1)). Consequently, the overall westward drift (forced by the magnetic curvature and cross-latitudinal gradient) is reduced by the eastward component – exerted in regions with a negative azimuthal magnetic gradient. Furthermore, the electric field (induced by the charge separation of impending particles) is significantly reduced in these regions. Finally, the number of particles expelled outside the magnetic trap (due to the $(E \times B)/B^2$ electric drift) is much less. More precisely, only a few of them have a "chance" to be lost in the atmosphere in said regions.

Oppositely, in regions with positive azimuthal geomagnetic gradients, the southward drift component changes slightly in the direction, but not the amplitude of the westward drift, impelled by the magnetic curvature and latitudinal gradient. Consequently, in these regions, the induced electric field – resulted from the charge separation of arriving particles – is much stronger. It will intensively expel the charged particles outside the magnetic trap through the imposed $(E \times B)/B^2$ drift. Furthermore, these particles interact with the atmospheric molecules creating secondary electrons, ions, and nuclear products, giving rise to the ionization of the lower atmosphere.

3.3 Hemispherical asymmetry of geomagnetic non-dipolar field and its influence on particle precipitation in Earth's atmosphere

The confinement of any particle in the gradient magnetic field B depends on the ratio between the maximum field strength B_{max} in the polar regions (where the backward reflection of trapped particles occurs) and the equatorial magnetic field strength B_0, i.e.

$$\sin(\alpha) = \sin(\alpha_0) \cdot \sqrt{\frac{B_{\max}}{B_0}} \tag{3}$$

where the angle α_0 between velocity vector of arriving particle and corresponding magnetic line in the equatorial region, is known as an equatorial *pitch* angle, and α is the continuously changing pitch angle, when particle is moving along the magnetic field line. Thus α increases with particles' movement toward the pole, due to the reduction of filed aligned component of particles velocity, and increase of its velocity in a direction perpendicular to geomagnetic field line (refer to formula (4) and **Figure 4**), are decreases when particle is moving toward the equator.

$$\alpha = 2\pi\left(\frac{v_{II}}{v_{\perp}}\right) \cdot r_B, \quad \text{where } r_B = \frac{m}{qB} v_{\perp} \tag{4}$$

Any particle is assumed trapped by the magnetic field, when the angle α becomes greater than $\frac{\pi}{2}$, because at this point – known as a magnetic mirror – the particle reverses its direction of movement, remaining confined by the magnetic field line. Formula (3) shows that particles approaching Earth's magnetosphere at very small angles could not exceed the pith angle $\frac{\pi}{2}$, and when enter the mirror point these particles are "lost" in the atmosphere. The minimum value of angle α_{0m} (for which the maximum magnetic field is still able to reflect particles) is called *loss cone*. If a particle arrives at angle lower than the solid angle defined by α_{0m}, it will be lost in the ambient atmosphere on its motion along the magnetic field line. Formula (3) shows also that the efficiency of magnetic mirror to reflect charged particles does not depend

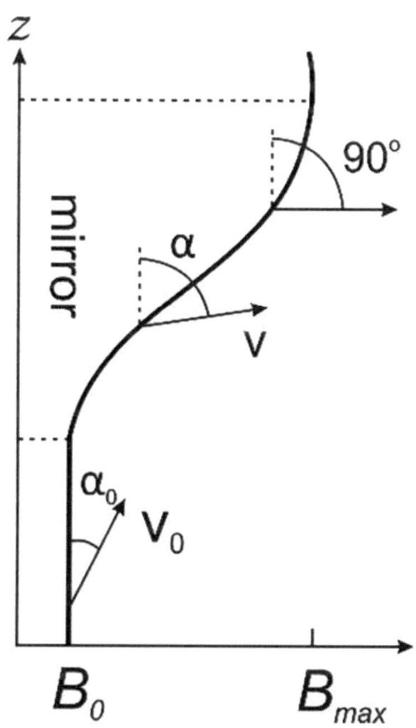

Figure 4.
Orientation of the particle's velocity vector, with respect to the equatorial magnetic field B_o, and changing particles pitch angle α (from α_o at the equator, to 90 degrees at magnetic mirror point).

neither on the particles speed, nor on their charge and mass (in the guiding center approximation, known also as adiabatic approximation).

The geomagnetic field near the poles is stronger in the Southern Hemisphere, compared to those in the Northern Hemisphere. Consequently, in the case of isotropic particles' flux arriving at magnetopause – almost every third particle will be confined in the Southern Hemisphere, while in the Northern Hemisphere less than ¼ of all arriving particles are trapped, because of its larger loss cone [51]. This means that some of particles confined in the Southern Hemisphere could not be held by the weaker geomagnetic field in the Northern Hemisphere. The expected result is – more particles precipitating in the Northern Hemisphere.

3.4 Regener-Pfotzer maximum and its influence on the lower stratospheric chemistry

Energetic particles penetrating deeper in the atmosphere create showers of secondary particles, produced from their interaction with atmospheric molecules – the deeper the penetration is, the wider the showers are. In the lower stratosphere, the number of secondary products dramatically increases, becoming maximal at a certain level. This level is known as a Regener-Pfotzer maximum. Beneath it, the concentration of secondary ions and electrons decreases again.

The longitudinal geomagnetic gradient and hemispherical asymmetry of geomagnetic field determine the uneven distribution of geomagnetically trapped particles' precipitation over the globe (refer to Subsections 3.2 and 3.3). Existence of such an effect is illustrated in [52].

3.4.1 Ozone formation in the lower stratosphere

For almost a century –since the creation of the theory about ozone production in the upper atmosphere by Sydney Chapman [53] – the single source of stratospheric ozone is believed to be the photo-dissociation of molecular oxygen by solar ultraviolet radiation. Recently it has been shown that in the dry lowermost stratosphere the lower-energy electrons in the Regener-Pfotzer maximum initiate ion-molecular reactions producing ozone [54].

The mean energy of electrons in the Regener-Pfotzer max ($\sim$35 eV [55]) is not sufficient to break the molecular bounds of the major atmospheric constituents. It is, however enough to ionize the molecular oxygen (Reaction (5)). The oxygen cation interacts furthermore with neutral oxygen molecule, producing a tetra-oxygen ion O_4^+ [56, 57], (see Reaction (6)).

Being very unstable, this oxygen complex rapidly dissociates into two different channels [57]. The first channel (7) produces O_3^+ and O, while the second one restores the O_2^+ ions (8). The weakly bonded O_3^+ molecule easily dissociates or exchanges its charge with O_2, yielding a neutral ozone. Most efficient, however, appears to be the dissociative recombination of ozone cation O_3^+ to three oxygen atoms, occurring in 94% of all cases [58], in prevailing conditions typical for the lower stratosphere (i.e. ground state ozone cations and lower energetic electrons).

$$O_2 + e^- \rightarrow O_2^+ + 2e^- + 12.07\ \text{eV} \tag{5}$$

$$O_2^+ + O_2 + M \rightarrow O_4^+ + M + 3.5\ \text{eV} \tag{6}$$

$$O_4^+ \rightarrow O_3^+ + O + 0.82\,\text{eV} \tag{7}$$

$$O_4^+ \rightarrow O_2^+ + O_2 + 1.26\,\text{eV} \tag{8}$$

$$O_3^+ + M \rightarrow O + O + O + \sim 0\,\text{eV} \tag{9}$$

$$O + O_2 + M \rightarrow O_3 + M \tag{10}$$

$$\text{Net}: 1O_4^+ \rightarrow 4O_3 \tag{11}$$

As a result, the dissociation of one O_4^+ molecule leads to the formation of four new O_3 molecules (reactions (7) and (9)), while reaction (8) and continuous ionization of O_2 by the atmospheric lower-energy electrons support a steady production of O_4^+ (more detailed analysis could be found in [51, 53]. Thus, the reactions (6)–(9) form an autocatalytic cycle for continuous O_3 production in the lower stratosphere. An absolutely necessary condition for the activation of autocatalytic ozone production is a dry atmosphere. Otherwise, water clusters of O_2^+ are formed instead of O_4^+ [59]. The maximum efficiency of this ozone-producing cycle should be expected near the level of the highest secondary ionization produced by GCRs, i.e. near the Regener–Pfotzer maximum.

3.4.2 Evidence for particles' influence on the lower stratospheric O_3 density

At middle and high latitudes, the Regener-Pfotzer maximum is placed well above the tropopause [60], which provides the necessary conditions for activation of the autocatalytic cycle of ozone production – i.e. a dry atmosphere and plenty of low energy electrons. As discussed in Sections 3.2 and 3.3, and shown in [52], the ionization in the Regener-Pfotzer maximum is unevenly distributed over the globe. Remind that an increased particles' flux is expected in regions of geomagnetic field strengthening. Consequently, if the autocatalytic production of ozone is significant, the longitudinal variations of the Regener-Pfotzer maximum ionization should be projected on the ozone profile.

Figure 5 presents a comparison between ozone profiles in regions with increasing and decreasing geomagnetic field, during solar minimum in 2009. Note that the O_3

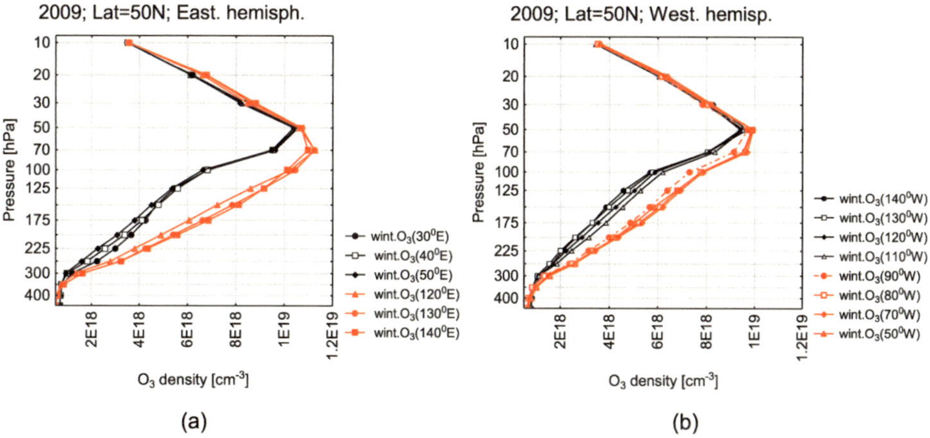

Figure 5.
Difference between ozone profiles in regions with positive (red curves) and negative (black curves) cross-longitudinal magnetic gradients; (a) for the Eastern hemisphere, and (b) for the Western one.

values beneath the peak ozone density are higher in regions with increasing geomagnetic field (i.e. the longitudinal sector 90–50°W in the Western Hemisphere and 120–140°E – in the Eastern one), relative to corresponding O_3 values in regions with a geomagnetic field weakening in the sectors: 140–110°W and 30–50°E.

The longitudinal variations in atmospheric ozone have been noticed long ago [61, 62]. The authors have suggested that this variability could be related to the planetary wave structure. However, the maximal amplitude of the stationary planetary waves is found at ~300 hPa [61], while the highest amplitude of O_3 longitudinal variations in ERA Interim reanalysis is placed near 150–70 hPa [51]. These and some other problems, e.g. [63, 64] suggest that other factor(s) (e.g. energetic particles) may have an important influence on the spatial and interannual variability of the extra-tropical near tropopause O_3.

In order to assess quantitatively the coupling between energetic particles precipitating in Earth's atmosphere and lower stratospheric ozone, as well as its spatial distribution, we have performed a cross-correlation analysis in a grid with 10° increments in latitude and longitude. Ground-based measurement of galactic cosmic rays (GCR) by neutron monitors, has been used as an indication of energetic particles flux. The Moscow record of GCR has been expanded backward in time by the paleore constructed GCR intensity [65]. The 11-year periodicity of GCR has been removed by moving averaging procedure with 22-year running window. The winter values of ozone at 70 hPa have been taken from ERA twentieth century reanalysis, covering the period 1900–2010. Data have been preliminarily smoothed by 11-year running window.

The map of ozone-GCR correlation is presented in **Figure 6** (colored shading). It is important to note that the map has been created from correlation coefficients, being preliminary weighted by the autocorrelation function of GCR with time lag corresponding to the delay of O_3 response to the GCR forcing. This procedure, which reduces correlation coefficients with longer time lags, allows a comparison of correlations with different time lags. The introduction of weighs for the lagged correlation coefficients is justified by the assumption that the effect of the applied forcing in a given moment of time decreases with moving away from this moment [66].

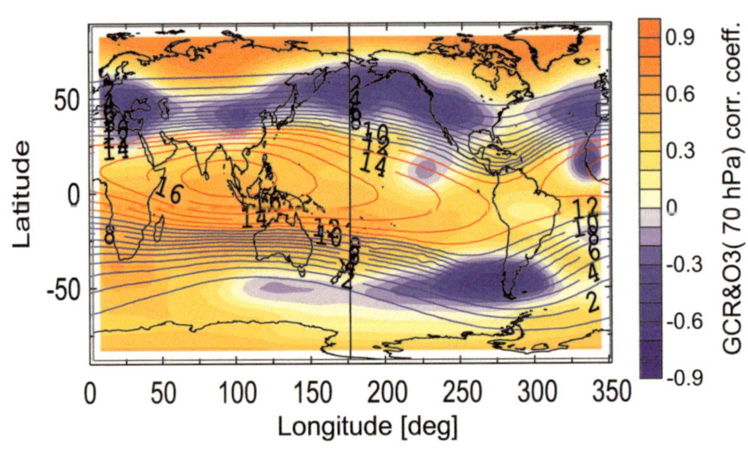

Figure 6.

Lag-corrected correlation map of GCR and O3 at 70 hPa (shading), compared with modeled effective vertical cut-of rigidity of geomagnetic field (courtesy to Boschini MJ, Della Torre S, Gervasi M., Grandi D, Rancoita PG: Http://www.mib.infn.it, and Bobik P, Kudela K: http://space.saske.sk).

Figure 6 shows that the ozone responds differently to particles' impact at different regions over the world – not only by amplitude but even by sign. Thus, at high latitudes and in the Indo-Pacific region, ozone varies synchronously with GCR. On the other hand, at the Northern Hemisphere extratropics and near the southernmost edge of Latin America, both variables covariate in antiphase – meaning that in these regions ozone increases with time.

Such heterogeneity in ozone response to particles' forcing could be attributed to the different origins of impacting particles. For example, the polar regions are vulnerable to the particles from interplanetary space, propagating along the open geomagnetic field lines. The long-term variations of these particles are modulated mainly by the interplanetary magnetic field in the heliosphere. The latitudes shielded by the closed geomagnetic field lines (i.e. the tropics and mid-latitudes) are accessible to very highly energetic particles (which are very few), and to the radiation trapped in the Van Allen radiation belts. The latter are subject to geomagnetic lensing (in the lowest part of their helical trajectories along the magnetic field lines) and asymmetrical precipitation in both hemispheres, due to the asymmetry of geomagnetic field (refer to Sections 3.2 and 3.3).

Figure 6 shows in addition the effective vertical cut-off rigidity of geomagnetic field (contours), with the values greater than 12 GV being colored in red. Note that the strongest GCR-O_3 correlation over the equatorial Indo-Pacific region fairly well coincides with the higher geomagnetic cut-off rigidity. Having in mind the centennial negative trend in GCR, the positive correlation coefficients indicate ozone depletion during the examined period (1900–2010). Consequently, the reduced ozone density could be attributed to the weaker particles' fluxes accessing the said region.

On the other side, the negative GCR-ozone correlation in extratropics suggests enhancement of ozone density near 70 hPa. This result indicates that particles confined in the outer radiation belt are involved in ozone production in the lower stratosphere. Powered by the solar wind, the population of this radiation belt is highly variable [45], reflecting the changes in solar activity. The examined period is characterized by enhanced solar activity, which appears to be projected on the extratropical latitudes as enhanced ozone density at 70 hPa – due to the enhanced particles' population in the outer radiation belt.

The positive GCR-O_3 correlation at polar latitudes suggests a centennial ozone depletion, which corresponds to the decreased flux of GCR, modulated itself by the stronger interplanetary magnetic field in the heliosphere during the twentieth century [51].

The centennial changes in ozone mixing ratio at 70 hPa, between the first decades of twenty-first and twentieth centuries, is presented in **Figure 7**. Note that ozone changes deduced from the correlation map in **Figure 6** fairly well corresponds to the observed changes of ozone at 70 hPa.

4. Ozone as a mediator of geomagnetic field influence on climatic variables

The sensitivity of atmospheric temperature profiles and climate to the ozone density (particularly near the tropopause) has been noticed long ago [67–71], etc. The detected synchronization between the spatial and temporal variability of particles' flux reaching the ground, and the lower stratospheric ozone, is a hint that ozone could serve as a mediator of the geomagnetic field-energetic particles' influence on climatic

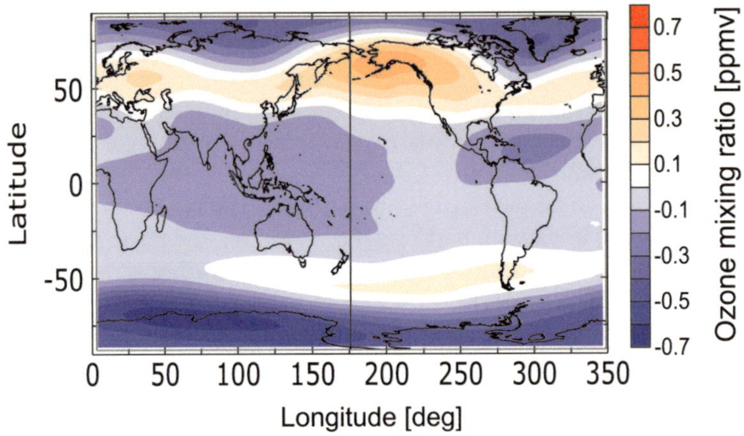

Figure 7.
Spatial distribution of centennial ozone changes between the first decades of twenty-first and twentieth centuries.

variables (i.e. temperature, pressure, etc.) [72]. The following section throws some more light on this problem.

4.1 Ozone imprints on climatic variables

4.1.1 Hemispherical and longitudinal asymmetries of ozone-temperature covariance

The potential synchronization between ozone at 70 hPa and near-surface temperature variability, within the period 1900–2010, is examined by the use of lagged cross-correlation analysis. The leading role of winter ozone in the ozone-temperature correlation, have been analyzed in a spatial grid with 10° steps in latitude and longitude. The time series of both variables are taken from the monthly values provided by the ERA twentieth century reanalysis. The correlation map presented in **Figure 8** is created from the preliminary weighted correlation coefficients by the autocorrelation function of ozone, with lag corresponding to the time delay of temperature response – to account for the reduced weigh of covariances being away from the moment of applied forcing.

The most impressive of the results shown in **Figure 8** is the asymmetry of the temperature response to ozone variations. The positive O_3 –T_{2m} correlation coefficients – over Eurasia and the extratropical Pacific Ocean, unlike the overall negative correlation, require their explanation. In addition, the analysis of the long-term variations of ozone and temperature at 60°N latitude, and at longitudinal zones, 140 and 70°W (corresponding to the regions with positive and negative GCR-ozone correlation) are presented in **Figure 9**. It is important to note that the short-term variations are preliminarily filtered by data smoothing through 11-year running average procedure.

Figure 9 clarifies that the lower temperature trend of Eastern Asia corresponds to the higher ozone density at 70 hPa. Oppositely, the stronger warming in south-eastern Canada corresponds to a lower ozone density at 70 hPa, with a negative centennial trend. Examination of the global picture of twentieth century warming (presented in **Figure 10**) reveals that the "hot spots" of contemporary global warming (i.e. north-eastern Canada and Greenland, and the Southern Ocean – southward of Africa) correspond to the regions of negatively correlated ozone and temperature (refer to **Figure 8**). In opposite, the regions with in-phase co-varying ozone and temperature are characterized by weaker warming.

DOI: http://dx.doi.org/10.5772/intechopen.103695

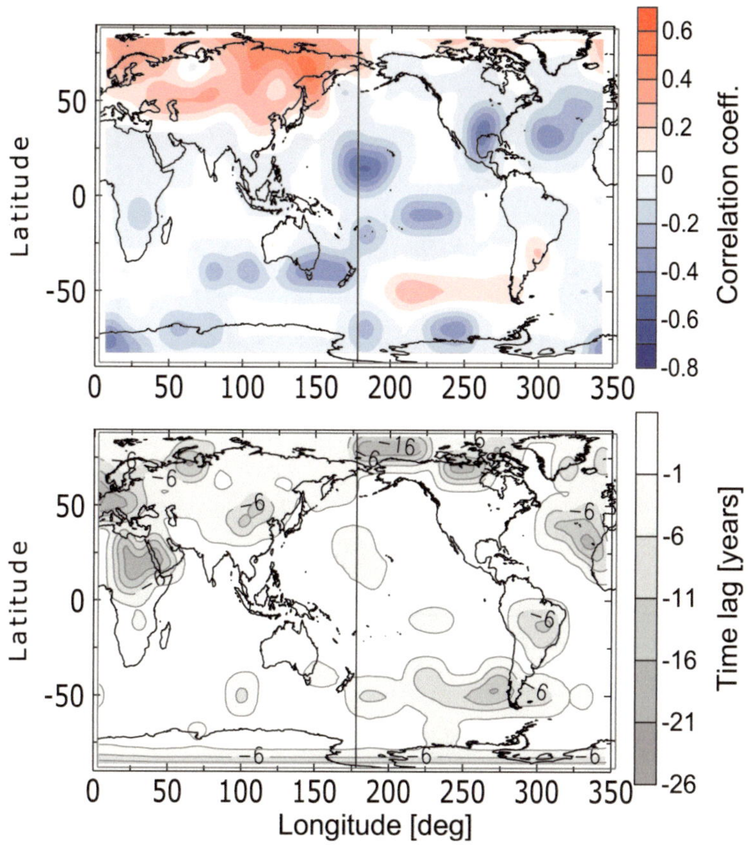

Figure 8.
(top) correlation map of winter ozone at 70 hPa and air surface temperature, calculated over the period 1900–2010; (bottom) time lag in years of temperature response following ozone changes.

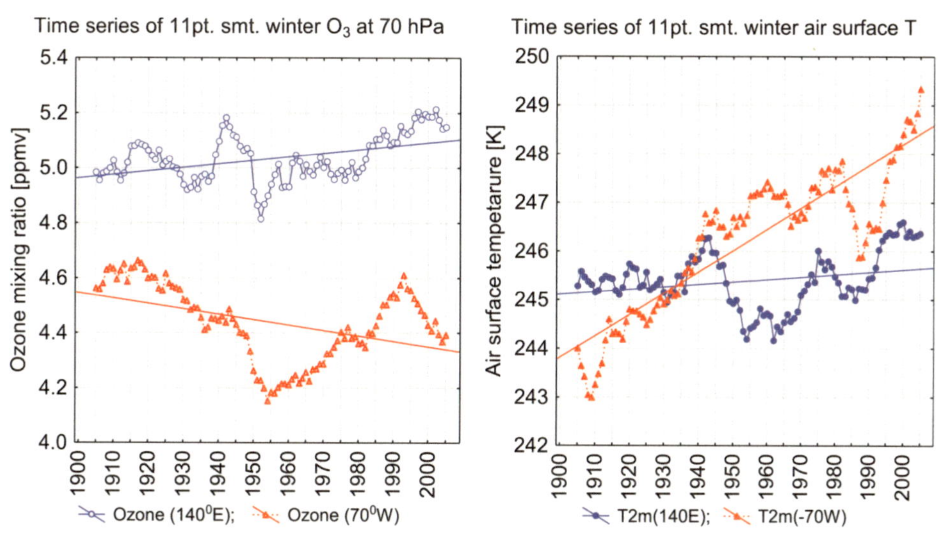

Figure 9.
(left) Time series of winter ozone at 70 hPa and 60°N latitude, obtained at Eastern (140°E longitude) and Western (70°W) longitude; (right) air surface temperature at the same latitude and longitudes.

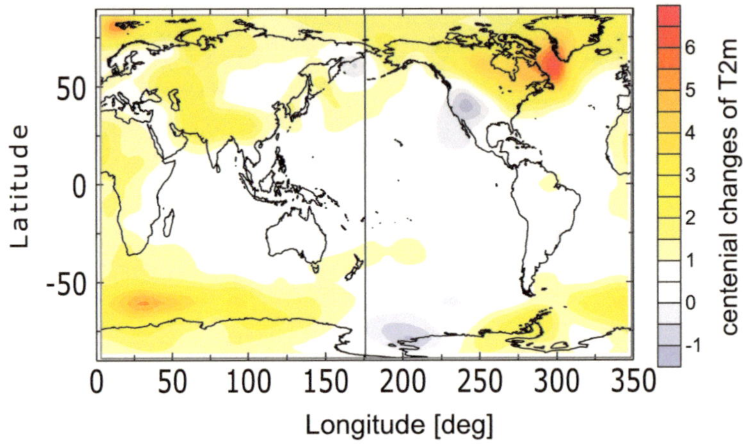

Figure 10.
Centennial changes of the air surface temperature between the first decades of twenty-first and twentieth centuries, derived from the ERA twentieth century reanalysis.

Conclusively, the above results indicate that the strongest warming during the twentieth century is observed in regions with reduced density of the lower stratospheric ozone.

4.1.2 Climatic modes and lower stratospheric ozone density

Climate variability is not homogeneous in space and is usually described as a combination of some "preferred" spatial regimes, called *modes*. In meteorology and climatology, the term *'mode'* is used to describe a spatial structure with at least two strongly connected centers of action [73]. The most famous of these spatial structures – known as *climatic modes* – affect weather and climate on different spatial and temporal scales. Most climatic *modes* are defined by statistical classifications of the observed variability of surface temperature, sea-level pressure, precipitations, etc. They could be a result of the action of fundamental physical processes such as the instability of the climatic mean flow, mesoscale interactions between the atmosphere and the ocean, etc. [74]. However, these statistical patterns may also be artifacts of nature, whereby they are not stable over long periods of time, or they may be statistical artifacts.

Although the spatial-temporal variations of climatic modes are extensively studied, the reasons for their occurrence and variability over time are not fully understood. Internal variations of the climate system are usually associated with the processes of energy exchange and redistribution between the planetary atmosphere and ocean. The huge heat capacity of the ocean is the reason for its inertia in response to short-time fluctuations of atmospheric variables, which transforms them into long-period variations of the ocean surface temperature. This understanding does explain the phase alteration, but it is not able to explain neither the various manifestations of climatic modes [75] nor their long-term changes.

Analysis of the spatial-temporal variability of GCR and ozone at 70 hPa reveals the important role of the latter in the formation of regional specificity of air surface temperature variability (refer to Subsection 4.1.1, or to [76]). Examination of the temporal synchronization between two of the most important *climatic modes* – North

Atlantic Oscillation (NAO) and El Niño Southern Oscillation (ENSO) –confirms the existence of statistical relation in the regions of *modes*' manifestation [77].

Figure 11 illustrates the projection of the long-term variations of ozone at 70 hPa on the NAO index (which describes the variability of the surface pressure between Azores and Iceland). The coupling between both variables has been estimated by the use of the lagged cross-correlation analysis between annual values of NAO index (smoothed by 5 points averaging) and winter ozone values at 70 hPa (smoothed by 11 points moving window). The stronger smoothing of ozone is due to its higher temporal variability. The leading factor (i.e. the "forcing") in calculated ozone-NAO variability is ozone. As in the previous case, the correlation coefficients have been preliminarily weighted (according to different delay of NAO response) with the ozone's autocorrelation function. The physical reasoning behind this weighting is that the memory of the climate system for the applied impact weakens with time. This suggests that the high correlation coefficients with a large delay are more or less random.

Figure 11 shows that the ozone's impact on the NAO climatic pattern fairly well coincides with both centers of action (Azores and Iceland) determining the phase of NAO mode. Unlike the previous results (stressing the leading role of the northern [78] or the southern part of NAO spatial structure [79]), **Figure 11** indicates that the variations of lower stratospheric ozone density can impact each center of action

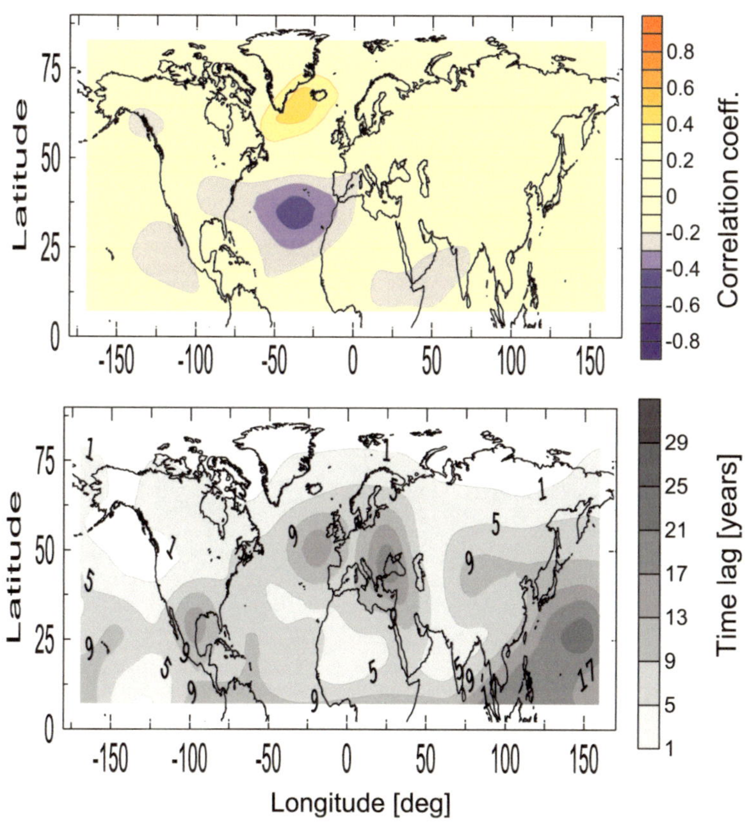

Figure 11.
(top) Cross-correlation maps of the winter lower stratospheric ozone and NAO index, calculated for the period 1900–2010; (bottom) time lag of NAO response in years.

(Azores or Icelandic), or simultaneously both of them – altering in such a way the phase of NAO mode [76].

Analysis of the time delay of NAO response to ozone changes shows that surface temperature near the Icelandic Low respond with a delay of 1–2 years. In the subtropical center of action, however, the atmospheric response is delayed approximately by a decade (see the bottom panel in **Figure 11**).

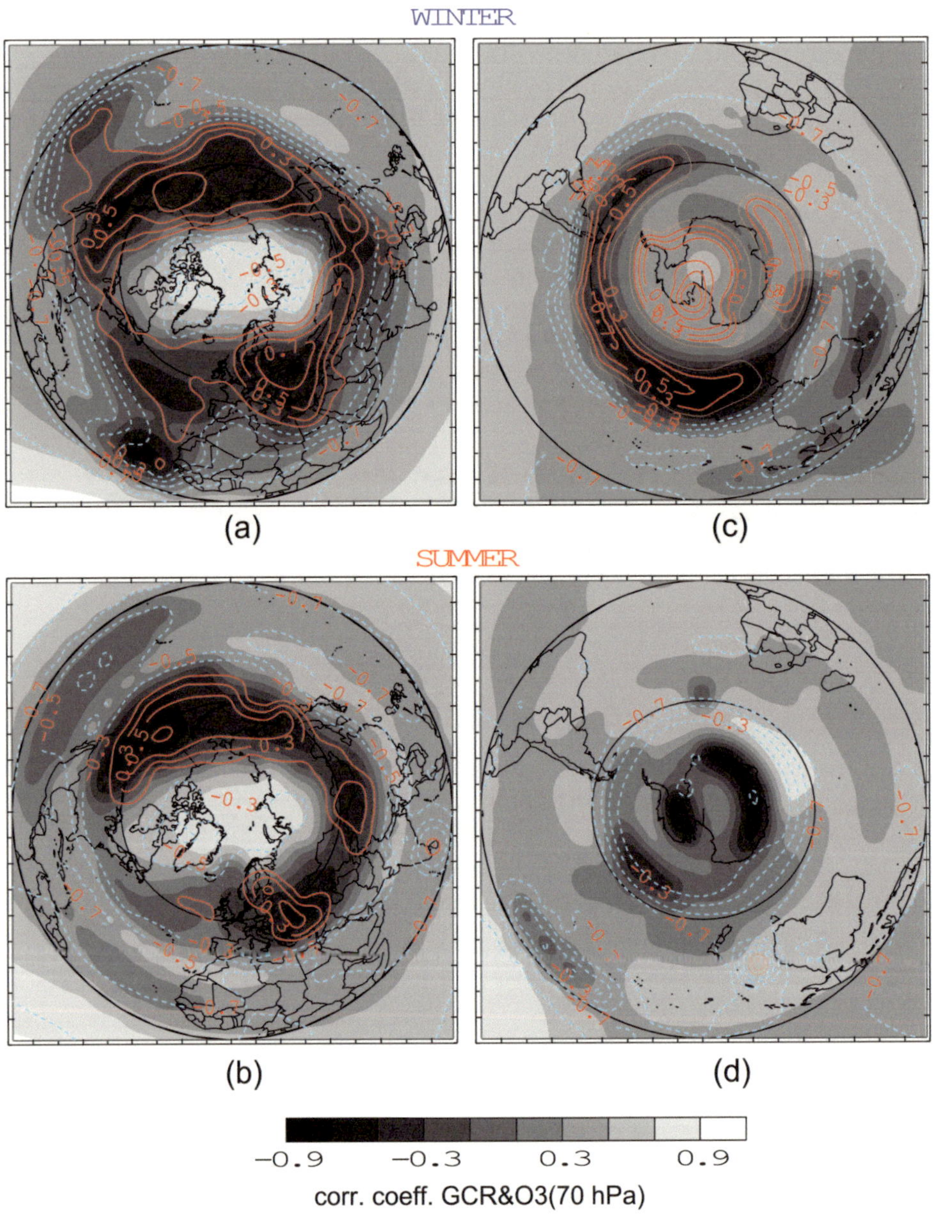

Figure 12.
Comparison of correlation maps of ozone at 70 hPa with GCRs (dark shading) and water vapor at 150 hPa (contours), for winter (a) and (c), and summer (b) and (d) panels.

4.2 Mechanism of ozone influence on climatic variables

Direct ozone influence on the surface temperature is quite small due to the mutually exclusive effect of stratospheric and tropospheric ozone in the planetary radiation balance [70]. Ozone's ability to absorb the incoming solar radiation (and to a lesser extent the longwave radiation emitted from the Earth), makes it a radiatively active gaz. The covariance between the near tropopause ozone and temperature has been noticed long ago [80, 81]. However, the tropopause temperature determines the moist adiabatic lapse rate and accordingly the static stability of the upper troposphere [82, 83], which in turn alters the humidity near the tropopause [51]. For example, ozone depletion cools the near tropopause region making the upper troposphere more unstable [82, 83]. The upward propagation of the more humid air masses from the lower atmospheric levels moistens the upper troposphere, and strengthens the greenhouse warming of the planet. The satellite measurements show that water vapor at these levels ensures 90% of the greenhouse warming of the total atmospheric humidity [84]. Consequently, ozone variability in the lower stratosphere is projected on the planetary surface through the modulation of the strength of greenhouse warming.

Figure 12, which compares the lag-corrected correlation maps of ozone mixing ratio at 70 hPa with: (i) GCR, and (ii) humidity at 150 hPa, is a good illustration of our hypothesis validity. Note that the latitudinal band of antiphase correlation between GCRs and ozone (dark shading), and in phase correlation between ozone and water vapor (red contours), coincide impressively well. In the Northern Hemisphere, this coincidence persists round the year, although being slightly reduced in summer season (compare panels (a) and (b) in **Figure 12**). In the winter Southern Hemisphere, the area of synchronous variations of GCR, ozone, and humidity is narrower and practically disappears in summer (**Figure 12d**). The results presented in **Figure 12** are a good indication that ozone–humidity variations, which are projected down to Earth's surface by the strengthening or weakening of the greenhouse effect, are actually related to GCR variability.

5. Conclusions

Historical and contemporary changes in climate system put a lot of questions, the answers to which are difficult. This motivates scientists from different branches to look for various factors with a potential influence on the climate system. Geomagnetic field is one of the proposed factors, due to the rendered multiple evidence for spatially or temporary co-varying geomagnetic field and climate, at different time scales. In this chapter, we clarify that hypothesized geomagnetic influence on climate could be reasonably explained through the mediation of energetic particles, propagating in Earth's atmosphere, and their influence on the ozone density in the lower stratosphere.

More specifically, the non-dipolar part of geomagnetic field creates irregularities in the spatial distribution of lower atmospheric ionization in the Regener-Pfotzer maximum [51]. The bulk of low-energy electrons and dry lower stratosphere favors activation of autocatalytic ozone production at these altitudes. Thus geomagnetic irregularities are projected on the ozone density near the tropopause. Being a radiatively active gas, the ozone itself affects the temperature and humidity in the tropopause region, altering in such a way the greenhouse effect and consequently – the near-surface temperature.

This chapter provides evidence for the validity of this chain of sequences, which gives an adequate explanation of hemispherical and longitudinal asymmetry of the lower stratospheric ozone distribution, regionality of climate change, formation of regional climate patterns, known as climatic modes, etc.

Acknowledgements

The authors are grateful to the project teams of ERA 20C and ERA Interim reanalyses, providing gridded data for ozone temperature and pressure, as well as to the Climatic research unit of University of East Anglia – for the data of NAO index. We are thankful to the National Centre for Environmental Information (NOAA) for providing IGRF model.

Funding

This research was funded by National Science Fund of Bulgaria, Contracts KP-06-N34/1/30-09-2020, and DN 14/1 from 11.12.2017.

Author details

Natalya Kilifarska[1*], Volodymyr Bakhmutov[2] and Galyna Melnyk[2]

1 Climate, Atmosphere and Water Research Institute, Bulgarian Academy of Sciences, Sofia, Bulgaria

2 Institute of Geophysics by S.I. Subbotin name, National Academy of Sciences of Ukraine, Kyiv, Ukraine

*Address all correspondence to: natalya_kilifarska@yahoo.co.uk

References

[1] Worm H-U. A link between geomagnetic reversals and events and glaciations. Earth and Planetary Science Letters. 1997;**147**:55-67. DOI: 10.1016/S0012-821X(97)00008-3

[2] Kitaba I, Hyodo M, Katoh S, Dettman DL, Sato H. Midlatitude cooling caused by geomagnetic field minimum during polarity reversal. Proceedings of the National Academy of Sciences. 2013; **110**:1215-1220. DOI: 10.1073/pnas.1213389110

[3] Nurgaliev DK. Solar activity, geomagnetic variations, and climate changes. Geomagnetism and Aeronomy. 1991;**31**:14-18

[4] Butchvarova V, Kovacheva M. European changes in the paleotemperature and Bulgarian archaeomagnetic data. Bulgarian Geophysical Journal. 1993;**19**:19-23

[5] Gallet Y, Genevey A, Fluteau F. Does Earth's magnetic field secular variation control centennial climate change? Earth and Planetary Science Letters. 2005;**236**:339-347. DOI: 10.1016/j.epsl.2005.04.045

[6] Vogt J, Zieger B, Glassmeier K-H, Stadelmann A, Kallenrode M-B, Sinnhuber M, et al. Energetic particles in the paleomagnetosphere: Reduced dipole configurations and quadrupolar contributions. Journal of Geophysical Research: Space Physics. 2007;**112**:A06216. DOI: 10.1029/2006JA012224

[7] Channell JET, Vigliotti L. The role of geomagnetic field intensity in Late Quaternary evolution of humans and large mammals. Reviews of Geophysics. 2019;**57**:709-738. DOI: 10.1029/2018RG000629

[8] Alken P, Thébault E, Beggan CD, et al. International geomagnetic reference field: The 13th generation international geomagnetic reference field: The thirteenth generation. Earth, Planets and Space. 2021;**73**(49). DOI: 10.1186/s40623-020-01288-x

[9] Mandea M, Korte M, Yau A, Petrovsky E, editors. Geomagnetism, Aeronomy and Space Weather: A Journey from the Earth's Core to the Sun (Special Publications of the International Union of Geodesy and Geophysics). Cambridge: Cambridge University Press; 2019. DOI: 10.1017/9781108290135

[10] Chapman S, Bartels J. Geomagnetism. Oxford: University Press; 1940. p. 1049

[11] Matsushita S. Solar quiet and lunar daily variation fields. In: Matsushita S, Campbell WH, editors. Physics of Geomagnetic Phenomena. Orlando, FL: Academic Press; 1967. pp. 302-424

[12] Campbell WH, Schiffmacher ER. Quiet ionospheric currents of the southern hemisphere derived from geomagnetic records. Journal of Geophysical Research. 1988;**93**(A2): 933-944. DOI: JA093iA02p00933

[13] Richmond AD. Ionospheric electrodynamics. In: Volland H, editor. Handbook of Atmospheric Electrodynamic. Vol. 2. Boca Raton, FL: CRC Press; 1995. pp. 249-290

[14] Richmond AD, Maute A. Ionospheric electrodynamics modelling. In: Huba J, Schunk R, Khazanov G, editors. Modelling the Ionosphere-Thermosphere System. Ch6. Chichester, UK: John Wiley; 2014. pp. 57-71. DOI: 10.1002/9781118704417

[15] Takeda M. Time variation of global geomagnetic Sq field in 1964 and 1980. Journal of Atmospheric and Solar – Terrestrial Physics. 1999;**61**:765-774

[16] Takeda M. Features of global geomagnetic Sq field from 1980 to 1990. Journal of Geophysical Research. 2002; **107**(A9):1252. DOI: 10.1029/2001JA 009210

[17] Yamazaki Y et al. An empirical model of the quiet daily geomagnetic field variation. Journal of Geophysical Research. 2011;**116**:A10312. DOI: 10.1029/2011JA016487

[18] Maeda H, Fujiwara M. Lunar ionospheric winds deduced from the dynamo theory of geomagnetic variations. Journal of Atmospheric and Terrestrial Physics. 1967;**29**:917-936

[19] Tarpley JD. The ionospheric wind dynamo–I: Lunar tide. Planetary and Space Science. 1970;**18**:1075-1090. DOI: 10.1016/0032-0633(70)90109-1

[20] Parkinson WD. Introduction to Geomagnetism. Edinburgh: Scottish Academic Press; 1983. p. 433

[21] Mishin V, Bazarzhapov A. Introduction to the Physics of the Earth's Magnetosphere: Basic Information, Problems and some Results of the MIT ISTP Group: (BSFP-2002, Presented Reports) [Internet]. 2002. Available from: http://space.msu.ru/irkutsk/mish in.htm [Accessed: January 7, 2022]

[22] Gonzalez WD, Tsurutani BT, Clúa de Gonzalez AL. Interplanetary origin of geomagnetic storms. Space Science Reviews. 1999;**88**:529-562. DOI: 10.1023/ A:1005160129098

[23] Hu Q, He W, Zhao L, Lu E. Configuration of a magnetic cloud from solar orbiter and wind spacecraft In-situ measurements front. Frontiers in Physics. 2021;**9**:706056. DOI: 10.3389/ fphy.2021.706056

[24] Fenrich FR, Luhmann JG. Geomagnetic response to magnetic clouds of different polarity. Geophysical Research Letters. 1998;**15**:2999-3002. DOI: 10.1029/98GL51180

[25] Gopalswamy N, Yashiro S, Xie H, Akiyama S, Mäkelä P. Properties and geoeffectiveness of magnetic clouds during solar cycles 23 and 24. Journal of Geophysical Reseach. Space Physics. 2015;**120**:9221-9245. DOI: 10.1002/ 2015JA021446

[26] Gonzalez WD, Echer E. A study on the peak Dst and peak negative Bz relationship during intense geomagnetic storms. Geophysical Research Letters. 2005;**32**:L18103. DOI: 10.1029/ 2005GL023486

[27] Laughlin LK, Turner NE, Mitchell EJ. Geoeffectiveness of CIR and CME events: Factors contributing to their differences. Journal of the South-eastern Association for Research in Astronomy. 2008;**2**:19-22

[28] Alves MV, Echer E, Gonzalez WD. Geoeffectiveness of corotating interaction regions as measured by Dst index. Journal of Geophysical Research: Space Physics. 2006;**111**:A07S05. DOI: 10.1029/2005JA011379

[29] Maksimenko O, Melnyk G. Global distribution of magnetic storm fields and relativistic particles fluxes. In: Kogan F, Powell A, Fedorov O, editors. Use of Satellite and In-Situ Data to Improve Sustainability. Dordrecht: Springer; 2011. pp. 295-304

[30] Yaremenko LN, Melnik GV. Magnetic storms on 15–16 July 2000 and

15 May 1997. Geofizicheskiy Zhurnal. 2005;**27**(5):874-882

[31] Joshi S, Rao KM. Influence of solar wind parameters on equatorial magnetic observatories during intense geomagnetic storms of the year 2015. Journal of Scientific Research. 2020; **12**(3):233-250. DOI: 10.3329/jsr.v12i3.42798

[32] Bhardwaj S, Khan PA, Atulkar R, Purohit PK. Variability of geomagnetic field with interplanetary magnetic field at low, mid and high latitudes. Journal of Scientific Research. 2018;**10**(2):133-144. DOI: 10.3329/jsr.v10i2.34509

[33] Mustel ER, Chertoprud VE, Kovedeliani VA. Comparison of changes of the field of surface air pressure in the periods of high and low geomagnetic activity. Astron. Zhurnal. (in Russian). 1977;**54**:682-697

[34] Bochníček J, Hejda P, Pýcha J. Comparison of solar and geomagnetic activity effects on the northern hemisphere weather changes. Studia Geophysica et Geodaetica. 2001;**45**: 133-154. DOI: 10.1023/A:1021812210921

[35] Bucha V, Bucha V Jr. Geomagnetic forcing of changes in climate and in the atmospheric circulation. Journal of Atmospheric and Solar – Terrestrial Physics. 1998;**60**:145

[36] Regi M, Redaelli G, Francia P, Lauretis MD. ULF geomagnetic activity effects on tropospheric temperature, specific humidity, and cloud cover in Antarctica, during 2003–2010. Journal of Geophysical Research. Atmosphere. 2017;**12**(2):6488-6501. DOI: 10.1002/2017JD027107

[37] Padgaonkar AD, Arora BR. Tropospheric vorticity responses to the solar magnetic sector structure and geomagnetic disturbances. Pure and Applied Geophysics. 1981;**119**:893

[38] Bochnicek J, Hejda P, Bucha V, Pycha J. Temperature/pressure deviations and prevailing winds in the NH winter troposphere in years of high and low geomagnetic activity. Annales de Geophysique. 1998;**16**:C887

[39] Martazinova VF, Bakhmutov VG, Chulkov IS. Geomagnetic activity and atmosphere circulation. Geofizicheskiy Zhurnal. 2004;**26**(1):96-108

[40] Maliniemi V, Asikainen T, Salminen A, Mursula K. Assessing North Atlantic winter climate response to geomagnetic activity and solar irradiance variability. Quarterly Journal of the Royal Meteorological Society. 2019;**145**: 3780-3789. DOI: 10.1002/qj.3657

[41] Danilov LJ. Effect of geomagnetic storms on ionospher and atmosphere. International Journal of Geomagnetism and Aeronomy. 2001;**2**(3):209-224

[42] Abel B, Thorne RM, Vampola AL. Solar cyclic behavior of trapped energetic electrons in earth's inner radiation belt. Journal of Geophysical Research. 1994;**99**:19427

[43] Lyons LR, Thorne RM. Equilibrium structure of radiation belt electrons. Journal of Geophysical Research. 1973; **78**:2142-2149. DOI: 10.1029/JA078i013p02142

[44] Baker DN, Kanekal SG, Li X, Monk SP, Goldstein J, Burch JL. An extreme distortion of the Van Allen belt arising from the "Halloween" solar storm in 2003. Nature. 2004;**432**(7019): 878-881. DOI: org/10.1038/nature03116

[45] Li W, Hudson MK. Earth's Van Allen radiation belts: From discovery to the Van Allen probes era. Journal of

Geophysical Research: Space Physics. 2019;**124**:8319-8351. DOI: 2018JA 025940

[46] Singer SF. Radiation belt' and trapped cosmic ray albedo. Physical Review Letters. 1958;**1**:171-173. DOI: 10.1103/PhysRevLett.1.171

[47] Hudson K, Elkington SR, Lyon JG, Marchenko VA, Roth I, Temerin M, et al. Simulations of radiation belt formation during storm sudden commencements. Journal of Geophysical Research. 1997; **102**(A7):14087-14102. DOI: org/10.1029/ 97JA03995

[48] Kress BT, Hudson MK, Slocum PL. Impulsive solar energetic ion trapping in the magnetosphere during geomagnetic storms. Geophysical Research Letters. 2005;**32**:L06108. DOI: 10.1029/ 2005GL022373

[49] Selesnick RS, Looper MD, Mewaldt RA. A theoretical model of the inner proton radiation belt. Space Weather. 2007;**5**:S04003. DOI: org/ 10.1029/2006SW000275

[50] Li X, Selesnick R, Schiller Q, Zhang K, Zhao H, Baker DN, et al. Measurement of electrons from albedo neutron decay and neutron density in near – Earth space. Nature. 2017;**552**:382-385. DOI: 10.1038/ nature24642

[51] Kilifarska NA, Bakhmutov VG, Melnyk GV. The hidden link between Earth's magnetic field and climate. Elsevier. 2020:228. ISBN: 979-0-12-819346-4

[52] Kilifarska N, Bojilova R. Geomagnetic focusing of cosmic rays in the lower atmosphere – Evidence and mechanism. Comptes Rendus de l'Academie Bulgare des Sciences. 2019; **72**(3):365-374. DOI: 10.7546/ CRABS.2019.03.11

[53] Chapman SA. A theory of upper-atmospheric ozone. Memoirs of the Royal Meteorological Society. 1930; **3**(26):103-125

[54] Kilifarska NA. An autocatalytic cycle for ozone production in the lower stratosphere initiated by galactic cosmic rays. Comptes Rendus de l'Academie Bulgare des Sciences. 2013;**66**(2): 243-225

[55] Porter HS, Jackman CH, Green AES. Efficiencies for production of atomic nitrogen and oxygen by relativistic proton impact in air. The Journal of Chemical Physics. 1976;**65**:154-167

[56] Cacace F, de Petris G, Rosi M, Troiani A. Formation of O_4^+ upon ionization of O_2: The role of isomeric O_4^+ complexes. Chemistry—A European Journal. 2002;**8**:3653-3659

[57] Cacace F, de Petris G, Troiani A. Experimental detection of tetraoxygen. Angewandte Chemie, International Edition. 2001;**40**:4062-4065

[58] Zhaunerchyk V, Geppert WD, Österdahl F, Larsson M, Thomas RD. Dissociative recombination dynamics of the ozone cation. Physical Review A. 2008; **77**:022704. DOI: 10.1103/ PhysRevA.77.022704

[59] Kilifarska NA. Hemispherical asymmetry of the lower stratospheric O_3 response to galactic cosmic rays forcing. ACS Earth and Space Chemistry. 2017;**1**: 80-88. DOI: 10.1021/acsearthspacechem. 6b00009

[60] Sarkar R, Chakrabarti SK, Pal PS, Bhowmick D, Bhattacharya A. Measurement of secondary cosmic ray intensity at Regener-Pfotzer height using low-cost weather balloons and its correlation with solar activity. Advances

in Space Research. 2017;**60**:991-998. DOI: 10.1016/j.asr.2017.05.014

[61] Hood LL, Zaff DA. Lower stratospheric stationary waves and the longitude dependence of ozone trends in winter. Journal of Geophysical Research-Atmospheres. 1995;**100**:25791-25800. DOI: 10.1029/95JD01943

[62] Peters DHW, Gabriel DHW, Entzian G. Longitude-dependent decadal ozone changes and ozone trends in boreal winter months during 1960–2000. Annales de Geophysique. 2008;**26**: 1275-1286. DOI: 10.5194/angeo-26-1275-2008

[63] Pan L, Solomon S, Randel W, Lamarque JF, Hess P, Gille J, et al. Hemispheric asymmetries and seasonal variations of the lowermost stratospheric water vapor and ozone derived from SAGE II data. Journal of Geophysical Research – Atmospheres. 1997;**102**: 28177-28184. DOI: 10.1029/97JD02778

[64] Pan LL, Hintsa EJ, Stone EM, Weinstock EM, Randel WJ. The seasonal cycle of water vapor and saturation vapor mixing ratio in the extratropical lowermost stratosphere. Journal of Geophysical Research – Atmospheres. 2000;**105**:26519-26530. DOI: 10.1029/2000JD900401

[65] Usoskin IG et al. Cosmic Ray Intensity Reconstruction. IGBP PAGES/World Data Center for Paleoclimatology. Data Contribution Series 2008–013. Boulder, CO: NOAA/NCDC Paleoclimatology Program; 2008

[66] Kenny DA. Correlation and Causality. New York: John Wiley & Sons Inc.; 1979

[67] Manabe S, Strickler RF. Thermal equilibrium of the atmosphere with a convective adjustment. Journal of the Atmospheric Sciences. 1964;**21**:361-385. DOI: 10.1175/1520-0469(1964)021 <0361:TEOTAW>2.0.CO;2

[68] Manabe S, Wetherald RT. Thermal equilibrium of the atmosphere with a given distribution of relative humidity. Journal of the Atmospheric Sciences. 1967;**24**:241-259. DOI: 10.1175/1520-469(1967)024<0241:TEOTAW> 2.0.CO;2

[69] Ramanathan V, Callis LB, Boughner RE. Sensitivity of surface temperature and atmospheric temperature to perturbations in the stratospheric concentration of ozone and nitrogen dioxide. Journal of the Atmospheric Sciences;**33**:1092-1112. DOI: 10.1175/1520-0469(1976)033< 1092:SOSTAA>2.0.CO;2

[70] de Forster PMF, Shine KP. Radiative forcing and temperature trends from stratospheric ozone changes. Journal of Geophysical Research – Atmospheres. 1997;**102**:10841-10855. DOI: 10.1029/96JD03510

[71] Stuber N, Ponater M, Sausen R. Is the climate sensitivity to ozone perturbations enhanced by stratospheric water vapour feedback? Geophysical Research Letters. 2001;**28**:2887-2890. DOI: 10.1029/2001GL013000

[72] Kilifarska NA, Bakhmutov VG, Melnyk GV. The geomagnetic field's imprint on the twentieth century's climate variability. Geological Society, London, Special Publications. 2020; **497**(1):205-227. DOI: 10.1144/SP497-2019-38

[73] Wang G, Schimel D. Climate change, climate modes, and climate impacts. Annual Review of Environment and Resources. 2003;**28**:1-28. DOI: 10.1146/annurev.energy.28.050302.105444

[74] Wallace JM. North Atlantic oscillation/annular mode: Two paradigms– One phenomenon. Quarterly Journal of the Royal Meteorological Society. 2000;**126**:791-805. DOI: 10.1002/qj.49712656402

[75] Christensen JH et al. Climate phenomena and their relevance for future regional climate change. In: Stocker TF, Qin D, Plattner G-K, Tignor M, Allen SK, Boschung J, Nauels A, Xia Y, Bex V, Midgley PM, editors. Climate Change 2013: The Physical Science Basis. Contribution of Working Group I to the Fifth Assessment Report of the Intergovernmental Panel on Climate Change. Cambridge, UK and New York, NY: Cambridge University Press; 2013

[76] Velichkova T, Kilifarska N. Lower stratospheric ozone's influence on the NAO climatic mode. Comptes rendus de l'Acade'mie bulgare des Sciences. 2019; **72**(2):219-225. DOI: 10.7546/CRABS.2019.02.11

[77] Velichkova T, Kilifarska N. Inter-decadal variations of the Enso climatic mode and lower stratospheric ozone. Comptes rendus de l'Acad'emie bulgare des. Sciences. 2020;**73**(4):539-546

[78] Brugnara Y, Brönnimann S, Luterbacher J, Rozanov E. Influence of the sunspot cycle on the northern hemisphere wintertime circulation from long upper air data sets. Atmospheric Chemistry and Physics. 2013;**13**:6275-6288

[79] Gray LJ, Woollings TJ, Andrews M, Knight J. Eleven-year solar cycle signal in the NAO and Atlantic/European blocking. Quarterly Journal of the Royal Meteorological Society. 2016;**142**: 1890-1903. DOI: 10.1002/qj.2782

[80] de Forster PMF, Tourpali K. Effect of tropopause height changes on the calculation of ozone trends and their radiative forcing. Journal of Geophysical Research – Atmospheres. 2001;**106**: 12241-12251. DOI: 10.1029/2000JD900813

[81] Seidel DJ, Randel WJ. Variability and trends in the global tropopause estimated from radiosonde data. Journal of Geophysical Research. Atmospheres. 2006;**111**(D21):D21101. DOI: 10.1029/2006JD007363

[82] North GR, Eruhimova TL. Atmospheric Thermodynamics: Elementary Physics and Chemistry. 1st ed. Cambridge: University Press; 2009. ISBN-13: 978-0521899635

[83] Young JA. Static stability. In: Holton JR, Curry JA, Pyle JA, editors. Encyclopedia of Atmospheric Sciences. 2nd ed. Vol. 5. Amsterdam: Academic Press; 2015. pp. 2114-2120

[84] Inamdar AK, Ramanathan V, Loeb NG. Satellite observations of the water vapour greenhouse effect and column long wave cooling rates: Relative roles of the continuum and vibration-rotation to pure rotation bands. Journal of Geophysical Research – Atmospheres. 2004;**109**:D06104. DOI: 10.1029/2003JD003980

Chapter 3

Magnetospheric Current Systems and the Polar Cap Index

Peter Stauning

Abstract

The transpolar convection of plasma and embedded magnetic fields generated by the solar wind interaction with the magnetosphere can be characterized by the polar cap (PC) indices, PCN (North) and PCS (South). These indices are derived from polar magnetic variations and calibrated with respect to the solar wind merging electric field (coupling function), E_M, considered to control the entry of solar wind energy into the magnetosphere providing power to disturbance processes such as magnetic storms, auroral substorms, and upper atmosphere heating. Thus, the PC indices could be used to quantify the solar wind intensities for solar-terrestrial research and to survey the entry of solar wind energy for space weather monitoring. The closest relations between PC indices and the geomagnetic disturbance processes are obtained by using the dual polar cap PCC indices built from the positive values of PCN and PCS. The present work demonstrates that the transpolar convection processes characterized by the PCC indices are closely related to the intensities of auroral electrojet currents, to substorm occurrences, and to the building of magnetospheric ring currents in the equatorial region at 4–6 earth radii distance.

Keywords: solar wind/magnetosphere interactions, polar cap index, auroral electrojet currents, magnetospheric ring currents, magnetic storms and substorms, space weather forecasting.

1. Introduction

The magnetosphere comprises a number of current systems contained within the geomagnetic bubble carved out in the solar wind flow of tenuous ionized gasses flowing from the solar surface carrying solar magnetic fields out into space.

Dungey [1] formulated the concept of magnetic merging processes taking place at the front of the magnetosphere between the interplanetary magnetic field (IMF), when southward oriented, and the geomagnetic field, followed by the draping of the combined solar and geomagnetic fields and associated ionized plasma over the poles creating an elongated magnetospheric structure. In the extended magnetospheric tail region, the geomagnetic field from the northern and southern hemisphere would reconnect releasing the solar magnetic fields. The restored geomagnetic field would then be convected sunward at lower latitudes to resume merging with the solar wind field at the front of the magnetosphere.

The high-latitude antisunward ionospheric and magnetospheric plasma drift across the polar cap (PC) and the return flow in the sunward motion along dawn and dusk auroral latitudes generate the two-cell "forward convection" patterns, later termed DP2 (Polar Disturbance type 2) by Obayashi [2], Nishida [3], and Nishida and Maezawa [4]. Subsequently, Dungey [5] extended his model to include cases where IMF is northward (NBZ conditions), which in stronger cases would reverse the convection patterns in the central polar cap and generate sunward transpolar "reverse convection" plasma flow later termed DP3 (Polar Disturbance type 3) possibly inside a residual two-cell forward convection system. Although many details have been added later [6], these solar wind-magnetosphere interaction models still prevail now, 60 years later. The strictly southward or northward IMF directions in the idealized models have been extended to all IMF directions while retaining the basic features of northward versus southward IMF orientation.

In addition to the magnetopause currents (MPC) marking the interface between the solar wind and geomagnetic space, the magnetosphere comprises the polar cap transpolar currents and the auroral current systems both intimately connected to the plasma convection, the tail current sheet connecting the magnetopause flanks, and the ring currents of ions encircling the Earth at middle, low, and equatorial latitudes at distances of 4–6 earth radii. It is the primary objective of the present contribution to demonstrate that these magnetospheric current systems are closely interrelated in terms of the polar cap (PC) indices, notably the dual polar cap PCC indices.

The polar cap indices, PCN (North) and PCS (South), based on magnetic data recorded at the central polar cap observatories in Qaanaaq (Thule) in Greenland and Vostok in Antarctica, respectively, were developed from the initial concept by Fairfield [7] through the pioneering works of Kuznetsov and Troshichev [8], Troshichev and Andrezen [9], and Troshichev et al. [10]. Further PC index developments were made by Vennerstrøm [11], Troshichev et al. [12, 13], Stauning et al. [14], and Stauning [15–20].

To derive PC index values, magnetic variations related to the transpolar convection of plasma and magnetic fields are calibrated against the values of the merging electric field (coupling function), E_M (=E_{KL}, [21]), derived from parameters in the impinging solar wind. The calibration parameters are based on the statistical processing of solar wind and geomagnetic data throughout an epoch of accumulated values. Through their association with E_M, the PC indices represent the merging processes between the solar wind magnetic fields extending from the Sun and the terrestrial magnetic fields at the magnetospheric boundaries and could be considered representative of the energy transfer from the solar wind to the magnetosphere. This energy may be temporarily stored in the magnetospheric tail configuration to be dissipated in processes such as auroral substorms, upper atmosphere heating, and ring current enhancements. In further developments, interactions between the solar wind and polar cap convection processes include the effects of the related field-aligned current systems and also the consideration of reconnection processes at the nightside [22].

Janzhura et al. [23] have used the PC indices in substorm studies to predict the duration of the growth phase at substorm developments. For isolated events, they estimated that substorm onset would occur as the PC index level reached ~2 mV/m. From the investigations of a large number of substorms, Troshichev et al. [24] concluded that substorm onset was likely to happen when the PC index starting from a low level exceeded 1.5 ± 0.5 mV/m.

Troshichev et al. [25] and Troshichev and Sormakov [26] have used PC indices to predict the maximum intensities (SYM-H minima) during geomagnetic storms. In the studies of geomagnetic storms by Stauning et al. [27] and Stauning [28, 29], the PC indices have been implemented in gradient source functions used to predict the development of ring current intensities characterized by Dst index values.

Among important applications of real-time PC indices are forecasts of strong substorms that may threaten power grids through their geomagnetically induced current (GIC) effects [30, 31]. An investigation of GIC-related high-voltage power line disturbances in Scandinavia [32] has demonstrated that the PC index values most often would remain at a high level for more than 2–3 h up to reported major power line cuts. The lengthy pre-event intervals, which are also reflected in the ring current indices [29], are most likely needed for enabling the merging processes at the front of the magnetosphere and subsequent transpolar convection characterized by the PC indices to load the tail configuration with enough energy to generate violent substorm events. The enhanced merging processes during extended pre-event intervals make the polar cap expand to enable substorm activity reaching subauroral latitudes, where important power grids reside. According to these investigations, PC index levels above ~10 mV/m maintained throughout more than 1 h should cause alert for subauroral power grids [32, 33].

Strong auroral currents in the polar ionosphere characterized by large PC index values may cause heating of the upper atmosphere, which would then expand to cause anomalies in satellite orbits. The ring current intensities characterized by the Dst indices, which are related to PC index values, have been associated with further space weather effects such as spacecraft charging. The resulting electrostatic discharges in spacecraft structures may cause harmful anomalies in satellite electronic systems [34].

The report ISO/TR23989:2020 [35] issued by the authoritative Technical Committee of the International Organization for Standardization (ISO) for the natural and artificial space environment discusses the operational estimation of the solar wind energy input into the Earth's magnetosphere. The report aims at providing guidelines for the use of operative ground-based information on the polar cap magnetic activity defined by the PC indices. The report notes: "*The solar wind energy incoming into the magnetosphere predetermines development of the magnetospheric disturbances: magnetic storms and substorms. Magnetospheric disturbances include a wide range of phenomena and processes directly affecting human activity, such as satellite damage, radiation hazards for astronauts and airline passengers, telecommunication problems, outages of power and electronic systems, effects in the atmospheric processes, and impact on human health.*"

2. The polar cap (PC) index concept

The main purpose of the polar cap (PC) index concept is formulating a parameter that would quantify the transfer of energy from the solar wind to the magnetosphere to generate global geomagnetic disturbances such as magnetic storms and substorms. This makes the PC index fundamentally different from the auroral electrojet (AE) indices and further ground based magnetic indices such as the planetary disturbance index (Kp), and the ring current indices (Dst, SYM, and ASY), which represent the dissipation of the energy received from the solar wind. In the initial version by Troshichev and Andrezen [9], the polar cap index was derived directly from polar magnetic variations. However, this index type depends critically on daily and seasonal variations in ionospheric conductivities.

A major progress in the development of the PC index concept came with the work of Troshichev et al. [10] introducing the scaling of polar magnetic variations against the so-called merging or geoeffective electric fields in the formulation by Kan and Lee [21]. This energy coupling function is actually based on a theoretical concept using a particular projection of the electric field assumed to relate to the interface between two colliding magnetized plasma bodies. The merging electric field function holds the important solar wind parameters, the velocity, V_{SW}, and the interplanetary magnetic field (IMF) strength and orientation in the geocentric solar magnetospheric (GSM) representation. The energy coupling concept was initially developed by Akasofu [36–38] to provide the so-called epsilon (ε) parameter considered to be of major relevance for substorm developments and then modified by Kan and Lee [21] to provide a convenient relation between important solar wind parameters and the supply of energy to geomagnetic disturbances.

With the new PC index concept, the solar wind parameters (V_{SW} and IMF GSM-B_Y, GSM-B_Z) are replaced by polar cap horizontal magnetic variations ($\Delta \boldsymbol{F}_H$), which relate to polar current systems (horizontal and/or field-aligned) generated by the solar wind-magnetosphere interactions and measurable from ground. Thus, the estimates of solar wind energy input could be based on reliable and continuous ground-based observations. Furthermore, with the scaling against parameters in the solar wind, the new index would be independent (in principle) of local conditions such as variable ionospheric conductivities and observatory position within the polar caps.

3. Derivation of PC index values

The descriptions of the steps in the calculations of PC indices can be found elsewhere, for instance, in Troshichev et al. [12–14] or Stauning [17–20]. They are summarized here for convenience. The polar magnetic disturbance vector is defined by

$$\Delta \boldsymbol{F} = \boldsymbol{F} - \boldsymbol{F}_{RL} \quad (1)$$

where the reference level, $\boldsymbol{F}_{RL}$, is composed of the secularly varying component, $\boldsymbol{F}_{BL}$, and a daily varying term, $\boldsymbol{F}_{QDC}$, the quiet day curve (QDC), representing the daily magnetic variation during quiet conditions. Thus

$$\Delta \boldsymbol{F} = \boldsymbol{F} - \boldsymbol{F}_{BL} - \boldsymbol{F}_{QDC} \quad (2)$$

In order to focus on solar wind effects and reduce the influence from currents associated with localized features, such as density gradients, the horizontal magnetic variations, $\Delta \boldsymbol{F}$, of the recorded horizontal magnetic field vector series are projected to an "optimum direction" in space to provide the scalar projected variations, ΔF_{PROJ}.

The optimum direction is assumed to be perpendicular to the DP2 transpolar convection-related sunward equivalent currents and characterized by its varying angle, φ, with the dawn-dusk meridian.

The solar wind energy coupling function, E_{KL}, here named "merging electric field," E_M, because of its inherent dimension (mV/m) is defined as follows [21]:

$$E_M = V_{SW}\left(B_Y^2 + B_Z^2\right)^{1/2} \sin^2(\theta/2) : \theta = \arctan\left(B_Y/B_Z\right) \quad (3)$$

where V_{SW} is the solar wind velocity, B_Y and B_Z are the GSM components of the interplanetary magnetic field (IMF), and θ is the polar angle of the transverse IMF vector. In consequence of its role as an energy coupling function, the projected polar cap magnetic disturbances, ΔF_{PROJ}, are assumed being proportional to E_M:

$$\Delta F_{PROJ} = \alpha E_M + \beta \tag{4}$$

where α is the slope and β is the intercept parameter named from a graphical display of the relation between ΔF_{PROJ} and E_M.

The polar cap (PC) index is now defined by equivalence with E_M in the inverse relation of Eq. (4), i.e.,

$$PC = \left(\Delta F_{PROJ} - \beta\right)/\alpha : (\approx E_M) \tag{5}$$

With the relation in Eq. (5), the ΔF_{PROJ} scalar values are scaled to make the PC index equal (on the average) to values of E_M in the solar wind. The scaling of the polar cap magnetic disturbances to a quantity in the solar wind removes (in principle) the dependence on the daily and seasonally varying ionospheric conductivities and other local conditions, such as the location of the measuring polar magnetic observatory.

The projection angle for the projection of the horizontal magnetic variation vector in its geographic representation, (ΔF_X, ΔF_Y), in the (rotating) observatory frame at longitude, λ, to the optimum direction, φ, in space is defined by

$$V_{PROJ} = \text{Longitude } (\lambda) + \text{UTh} \cdot 15^\circ + \text{optimum direction angle } (\varphi) \tag{6}$$

where UTh is the UT time at the observatory in hours.

Thus, the projected magnetic variations could be expressed by

$$\Delta F_{PROJ} = \Delta F_X \cdot \sin\left(V_{PROJ}\right) \pm \Delta F_Y \cdot \cos\left(V_{PROJ}\right) : (+\text{for southern, –for northern hemisphere}) \tag{7}$$

The propagation delay, τ, between parameters at the reference location in space for the solar wind data and the location for related effects at the polar cap, and the values of the optimum angle, φ, are both estimated from searching optimum correlation between E_M and ΔF_{PROJ} [12–14, 17]. The correlation coefficient is usually around R = 0.75, while the delay from the magnetospheric bow shock nose (BSN) to recorded effects in the polar cap is close to τ = 20 min regardless of the rotating observatory positions within the magnetospheric polar cap. The delay varies little with seasonal and solar activity conditions.

The calibration parameters, the slope, α, and the intercept, β, are estimated for each moment of the day and year by linear regression between time-delay-adjusted samples of ΔF_{PROJ} and E_M using past data from an extended epoch, preferably a complete solar cycle [12–14, 17]. The regression parameters and the optimum angle values are usually derived as hourly values but interpolated and tabulated throughout the year at 1-min resolution. They are kept invariant over years unless a new index version is introduced.

Forward convection (DP2) patterns prevail during conditions, where the IMF B_Z component is negative or just small, and generate positive ΔF_{PROJ} values. The slope parameter (α) is positive and the intercept term (β) is relatively small. Hence, the PC index values, according to Eq. (5), are mostly positive. During positive (northward) and strong IMF B_Z (NBZ) conditions, reverse convection patterns (DP3) may emerge and generate negative ΔF_{PROJ} values, which, in turn, may generate negative PC index values.

The PCC (PC combined) indices are derived from the means of non-negative values of the PCN and PCS indices filling zeroes for negative index values [15]:

$$\text{PCC} = (\text{PCN if} > 0 \text{ or else } 0 + \text{PCS if} > 0 \text{ or else } 0)/2. \qquad (8)$$

Thus, the PCC index values are always non-negative like the merging electric field, E_M, used for the calibration of the individual polar cap indices. At negative PC index values in both the hemispheres, the global magnetic activity goes low like the PCC index values. However, there could still be local magnetic activity such as upper atmosphere auroral heating and reverse transpolar ionospheric convection. Positive PC index values in one hemisphere indicates unipolar solar wind energy entry and the generation of global magnetic disturbances in agreement with the positive PCC index levels even if the PC index for the other hemisphere is dominatingly negative and would generate negative PC index values by simple averaging of PCN and PCS values. Even at lengthy intervals of negative PCN and PCS values, the magnetosphere is not emptied of energy but usually enters a low-activity state.

4. Basic polar magnetic observations

The magnetic data used for the standard PCN indices are collected from Qaanaaq (THL) observatory in Greenland operated by the Danish Meteorological Institute (DMI), while the Danish Space Research Institute (DTU Space) operates the magnetic instruments and takes care of the data collection and processing. Data for the standard PCS indices are collected from the Antarctic Vostok observatory operated by the Arctic and Antarctic Research Institute (AARI) in St. Petersburg, while data for an alternative PCS index are collected at the French-Italian Dome Concordia (Dome-C) observatory [39, 40]. An alternative source for the PCN index is Alert observatory in Canada operated by the Canadian Energy and Mining administration. The characteristics of the four locations, including essential geomagnetic parameters based on the NASA VITMO application for 2021, are specified in **Table 1**.

Prior to their use in PC index calculations, the magnetic data are carefully examined. It is of major importance that the base-level values are correctly adjusted. As one of the important measures used to disclose possible problems, the monthly average X- and Y-component values are inspected. These values are derived as the means of measured values for all hours of the 5 quietest (QQ) days each month. These dates are

Observatory	Station	Latitude	Longitude	CGMlat	CGMlon	LT = 00	MLT = 00
Name	Acr.	Deg.	Deg.	Deg.	Deg.	UThrs	UThrs
Qaanaaq	THL	77.47	290.77	83.86	23.86	4.62	3.60
Alert	ALE	82.50	297.65	87.02	70.10	4.16	0.14
Dome-C	DMC	–75.25	124.17	–89.31	44.52	15.72	1.77
Vostok	VOS	–78.46	106.84	–84.04	56.64	16.88	0.95

Table 1.
Geographic and geomagnetic parameters at 100 km of altitude for selected stations.

defined by the International Service for Geomagnetic Indices (ISGI) available at http://isgi.unistra.fr. **Figure 1a** and **b** display the average values for the observed X and Y components from Qaanaaq (THL) and Vostok (VOS).

The average X- and Y-component values for Qaanaaq (THL) display smooth secular changes that are easily interpolated to create adequate baseline values throughout the displayed years. It is evident from **Figure 1b** that the definition of proper baseline values for Vostok presents challenges by the irregular variations and unexpected jumps. The base levels need comprehensive adjustments to remove irregular base-level changes and retain smooth secular variations only. Such adjustments are described (to some length) in [17].

The next step in the processing of the polar magnetic data is deriving the quiet daily variations, the quiet day curve (QDC), for each of the two horizontal vector

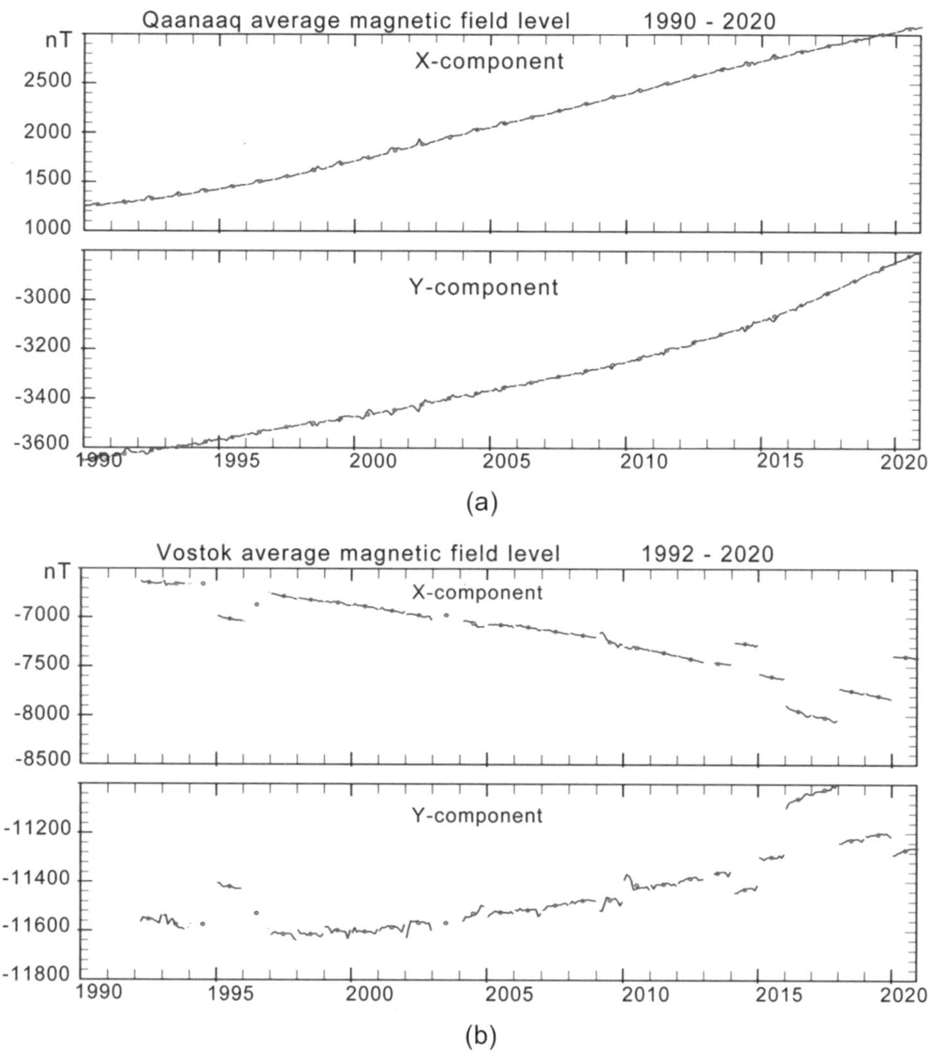

Figure 1.
Monthly (blue line) and yearly (red dots) average X- and Y-component values compiled throughout all hours of the 5 quietest days each month (http://isgi.unistra.fr). (a) Qaanaaq (THL) and (b) Vostok (VOS) (data from h ttps://intermagnet.org and http://www.wdc.bgs.ac.uk).

components. In the present work, the components are expressed in their (X,Y) representation, with X being the geographical northward component and Y the eastward component. Horizontal components in their geomagnetic representation or expressed by magnitude, *H*, and declination, *D*, could be used as well.

The definition of the level from which the magnetic variations should be measured is a controversial issue with different concept used by different PC index versions such as [12, 13, 41], or [42]. In the PC index version used, among others, in [17], the definition of the "solar rotation weighted" (SRW) reference-level construction published in [16] returns to the statements in [12, 13] with the vector formulation shown in Eq. (2) here and to the methods outlined in [43].

The essential point for the SRW method is deriving the reference level from quiet samples collected at conditions otherwise as close as possible to those prevailing at the day of interest. The factors of primary importance are:

i. Sample "quietness"

ii. Separation of the date of quiet samples from the QDC date

iii. Solar wind conditions (particularly IMF B_Y and V_{SW})

iv. Solar UV and X-ray illumination (based on solar radio flux F10.7 values)

For these factors, weight functions are defined to optimize the selection of samples for the QDC construction. For each hour of the day, observed hourly average values at corresponding hours within an extended interval (±40 days) are multiplied by the relevant weights, added, and then divided by the sum of weights to provide the hourly QDC value. Subsequently, the hourly QDC values are smoothed to remove irregular fluctuations and interpolated to provide any more detailed resolution, such as 1-min values, as required.

The weight function (i) for sample quietness is determined from the variability of 1-min data samples within the hour much like the technique used in [12, 13] and detailed in [43]. Two parameters are calculated on a vector basis. One is the maximum time derivative used to indicate the smoothness within the sample hour. The other is the average variance to define the slope of data values. Both the parameters need to take small values for the hourly sample to be considered "quiet" (flat and featureless display).

For an estimate on a statistical basis of further weight functions (ii)–(iv) related to the solar rotation, the factors of importance were subjected to autocovariance analyses versus separation between the date of the QDC and the dates of the quiet samples to be included in the construction of the QDC values. The autocovariance values should take large values to meet the condition that the quiet samples used to build the QDCs must represent conditions as close as possible to those prevailing at the day of interest. Particular attention should be given to variations in the IMF B_Y component associated with the solar wind sector structure.

The details of the autocovariance analysis are provided in [16]. The main results were, as could be expected, high autocorrelation values at nearby dates and also high values at dates displaced one full solar rotation of 27.4 days from the day of interest. On these days, the solar illumination and the solar wind conditions, such as the solar wind speeds and the solar sector structure (at two-sector structures), were similar to the prevailing conditions on a statistical basis. In between, at half a solar rotation, mixed IMF B_Y autocovariance results were found. In some cases, a local maximum was seen indicating the occurrences of four-sector solar wind structures. However, in most cases, the autocorrelation function had a deep minimum at half a solar rotation indicating two-sector structures. From these results, the above weight functions (iii) and (iv) were defined to take fixed values [16] making the QDC construction independent of parameters other than the measured polar magnetic variations.

Thus, at any time after initial 40 days of data collection, the relevant real-time QDC could be calculated, and after further 40 days of initial data collection, the final QDCs could be calculated for any day in the past. The hourly component averages and their quietness weight factors are fetched from their stored values, and their separation weight factors are found from the tabulated values. For each UT hour of the day, the hourly average component values within ±40 days are multiplied by the weight factors and summed up. The weight factors are summed up. The sum of weighted component hourly average values divided by the sums of weights defines for each hour the QDC value. The hourly sums of weights are quality factors for which alert limits could be set to caution against invalid values. The hourly QDC values are smoothed to remove fluctuations and then interpolated to provide the desired time resolution. The derived QDCs are routinely displayed in plots like those of **Figure 2a** and **b**.

In these diagrams for the X-components of the magnetic data from Qaanaaq (THL) and Vostok (VOS), there is a QDC curve for each day of the year. The daily QDC curves are drawn on top of each other in blue line for one month at a time. For day 1 (in black line), day 15 (yellow), and last day of the month (in red line), the QDCs are redrawn on top of the other QDCs. The additional curves provide an impression of the development of the QDCs throughout the month. The seasonal variations are very distinct in the developments of the monthly superposed QDCs with amplitude maxima at local summer. Most of the additional variability in the QDCs is caused by the IMF B_Y-related solar sector effects, which are then taken into account in the generation of appropriate daily QDCs. The displays of the Y-component QDCs are similar to the X-component displays.

The weighting over ±40 days makes the determination of the final QDC fairly insensitive to the intervals of missing data. Thus, the weighting technique allows the calculations of real-time QDCs with reduced accuracy from past data collected within –40 to 0 days (actual time) by simply ignoring the not yet available post-event samples without otherwise changing the ±40 days' calculation scheme. The QDCs could be improved gradually as new data arrive to be completed after passing +40 days with respect to the day of interest. Thus, with the predictable secular variations and the SRW-based QDC calculations defined here, there are seamless transitions between real-time and post-event QDC values.

5. PC index relations to the interplanetary merging electric field

With the methods for preparing the polar magnetic field variations defined in Section 4 and the formulas defined in Section 3, it is now possible to derive polar cap index values in post-event as well as in real-time versions. The PC indices are defined to match the merging electric field on the average throughout the reference epoch, which is 1997–2009 for Qaanaaq- and Vostok-based PCN and PCS values here. However, the question remains how well the PC indices match the merging electric field in specific cases and on different time scales.

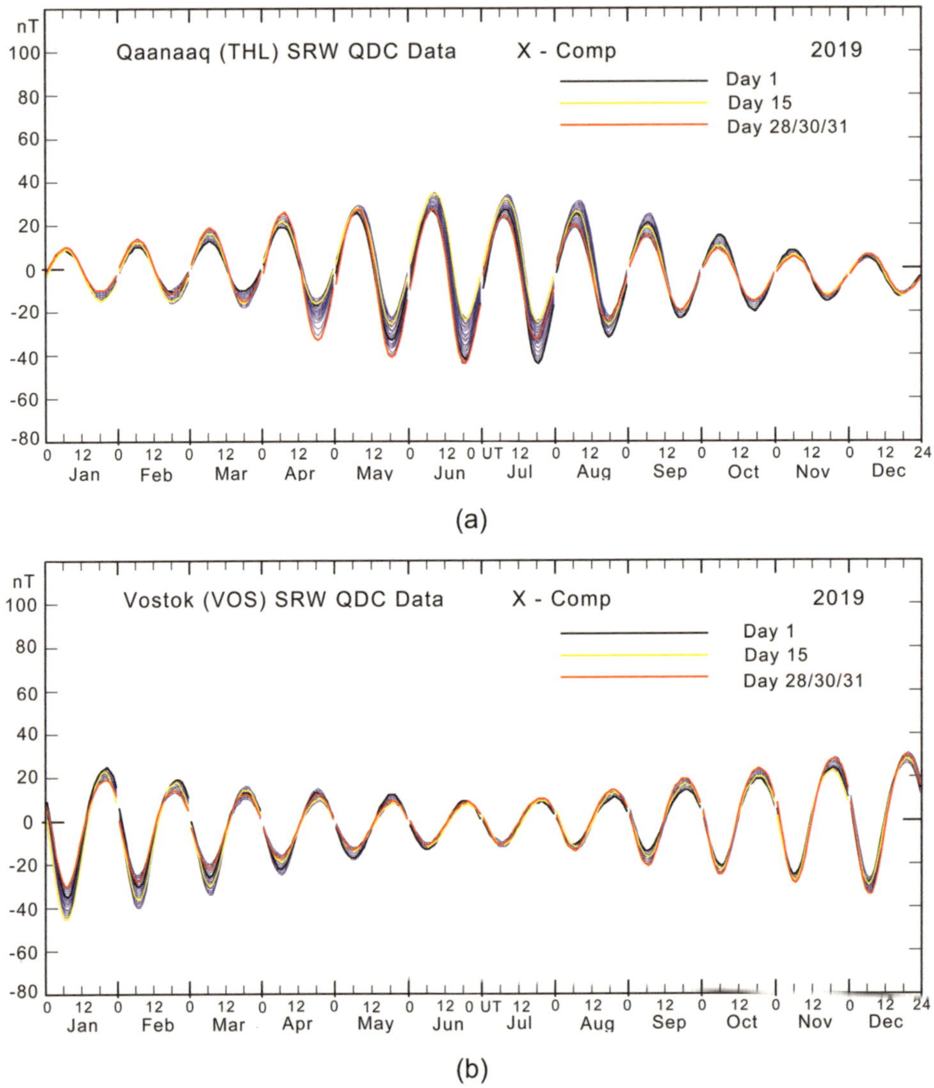

Figure 2.
QDC values for the X-components for Qaanaaq (THL) and Vostok (VOS) derived by SRW calculations. The monthly assemblies of QDCs are displayed in blue line. The variations on day 1, 15, and the last day of the month are superposed in black, yellow, and red lines, respectively.

In [29], the relations of the polar cap indices, PCN, PCS, and PCC, to the merging electric field, E_M [Eq. (3)], in the solar wind was investigated for the years from 1992 to 2018. The magnetic data supplied from INTERMAGNET (https://intermagnet.org) for Qaanaaq (THL) and Vostok were supplemented since 2009 by data from Dome-C observatory in Antarctica [39, 40]. The DMI2016 index calculation methods and coefficients [17] were used to derive index values.

Results from the correlations of PC index values in different versions with values of the merging electric field are displayed in **Figures 3** and **4**. The yearly average correlation coefficients are shown in **Figure 3**, where the coefficients for the correlation between PCN and E_M are displayed in blue line. The PCS-E_M correlation coefficients are depicted in red line, while the PCC-E_M correlation coefficients are shown by

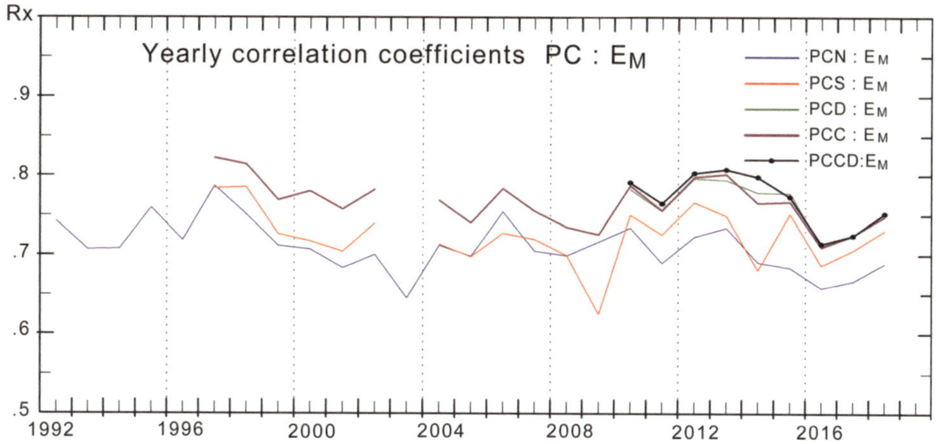

Figure 3.
Display of early averages of coefficients for the correlations between E_M *and PCN (blue line), PCS-Vostok (red), PCS (PCD)-Dome-C (green), PCC-(Qaanaaq-Vostok) (heavy magenta), and PCCD-(Qaanaaq-Dome-C) in heavy black line (after [29]).*

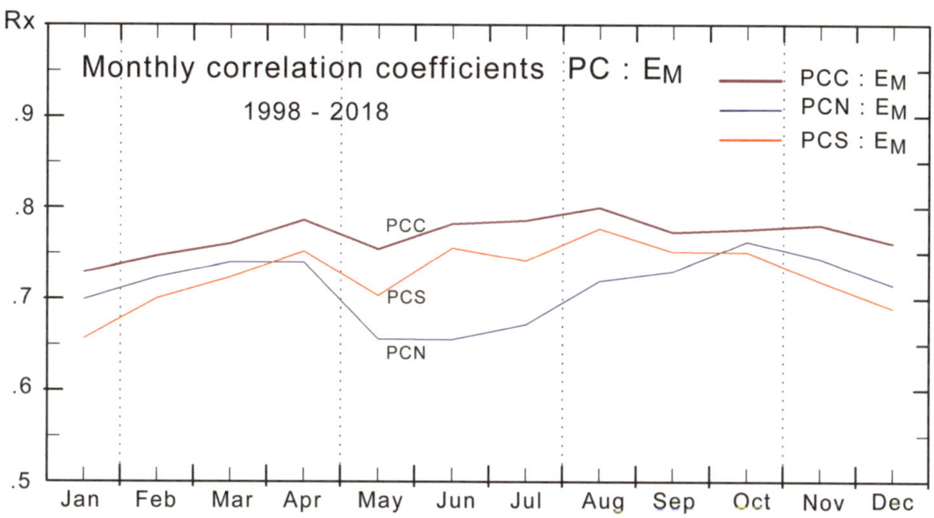

Figure 4.
Display of monthly average coefficients for the correlation between E_M *and PCN (blue line), PCS (red), and PCC (magenta) (from [29]).*

the heavy magenta line. In most cases, the PCC indices were derived from Qaanaaq-based PCN and Vostok-based PCS values. Data from Dome-C (DMC) observatory were used to derive an alternative PCS index, here denoted PCD, for the years 2012 and 2013, and further intervals where the Vostok data were incomplete. The correlation between E_M and PCD is displayed in green line, while the coefficients for correlation between E_M and PCC derived by using PCN and values of PCD substituted for PCS are displayed by the heavy black line named PCCD.

Figure 3 shows that the correlation between PCC and E_M is significantly higher than the correlation between E_M and either of PCN or PCS indices. It is also seen that the correlation between PCN and E_M is lower than the correlation between PCS and E_M with a few exceptions. It is seen from **Figure 3** that there is a tendency for decreasing correlations between E_M and either of the PC indices with time over the recent years. An in-depth investigation of this issue is beyond the scope of the present work.

Figure 4 displays the average correlation coefficients for calendar months January-December based on values from the years 1998–2018 with exception of 2003 void of PCS data and 2013 with incomplete PCS data. Values for Dome-C available since mid-2009 only have not been included in the display in **Figure 4**.

The correlations between PCN and E_M shown in **Figure 4** are clearly lower in the northern summer months, May–August, than for the rest of the year. Similarly, the correlations between PCS and E_M are clearly lower in the southern summer months, November-January, for Vostok than for the other seasons. Selecting the local winter index, PCW, by jumping between the PCN and PCS traces at equinoxes, improves correlation values while selecting the local summer index (PCU) reduces correlations. The overall correlation between PCC and E_M is clearly higher throughout all years and all seasons than the correlations between E_M and either of PCN, PCS, PCA (average of PCN and PCS), PCW, and PCU index versions.

In summary, the PCC index is the preferred index version to use for solar wind magnetosphere interaction studies and for investigations of global magnetic disturbances such as auroral current systems and magnetospheric ring current relations.

6. Relations to auroral current systems

Part of the more steady ionospheric auroral current system is related to the magnetospheric plasma convection patterns and could be considered to flow as horizontal Hall currents in the transition region between the field-aligned currents from the magnetopause regions flowing downward at the dawn-side, upward at the dusk-side, and field-aligned currents flowing from the ionospheric regions to the ring current regime, upward at dawn, downward at dusk [44–47].

The steady auroral currents are at times strongly intensified by dynamic substorm-related horizontal currents flowing between sheets of field-aligned currents originating from instabilities in the magnetospheric tail current structure [48, 49].

The present work shall focus on the relations between polar cap indices and indices describing auroral current intensities such as the auroral electrojet indices AL (lower envelope of negative magnetic bays) and AU (upper envelope of positive bays, [50, 51]) and the corresponding SuperMag indices, SML and SMU, based on a wider selection of observatories [52]. These relations shall be looked at in terms of linear correlations and regression between series of index data and examination of conditions for the rapid enhancements of the auroral electrojet intensities associated with the onsets of substorm events.

The intensities of the (equivalent) horizontal ionospheric currents in the dawn sector are generally best described by the AL (or SML) indices, while the intensities of the horizontal ionospheric currents in the dusk sector are described by the AU (SMU) indices. The substorm-related currents in the midnight sector are best represented by enhancements in the AL (SML) indices.

The examples of such relations are displayed in **Figure 5a** and **b** for 5-min samples of AL and SML and PCN, PCS, and PCC indices throughout the 4-day interval from 16 to 19 March 2015. It is seen that there are close correspondence between the (negative) auroral indices in green line and the PCN indices in blue, the PCS indices in red, and, in particular, the PCC indices in magenta line.

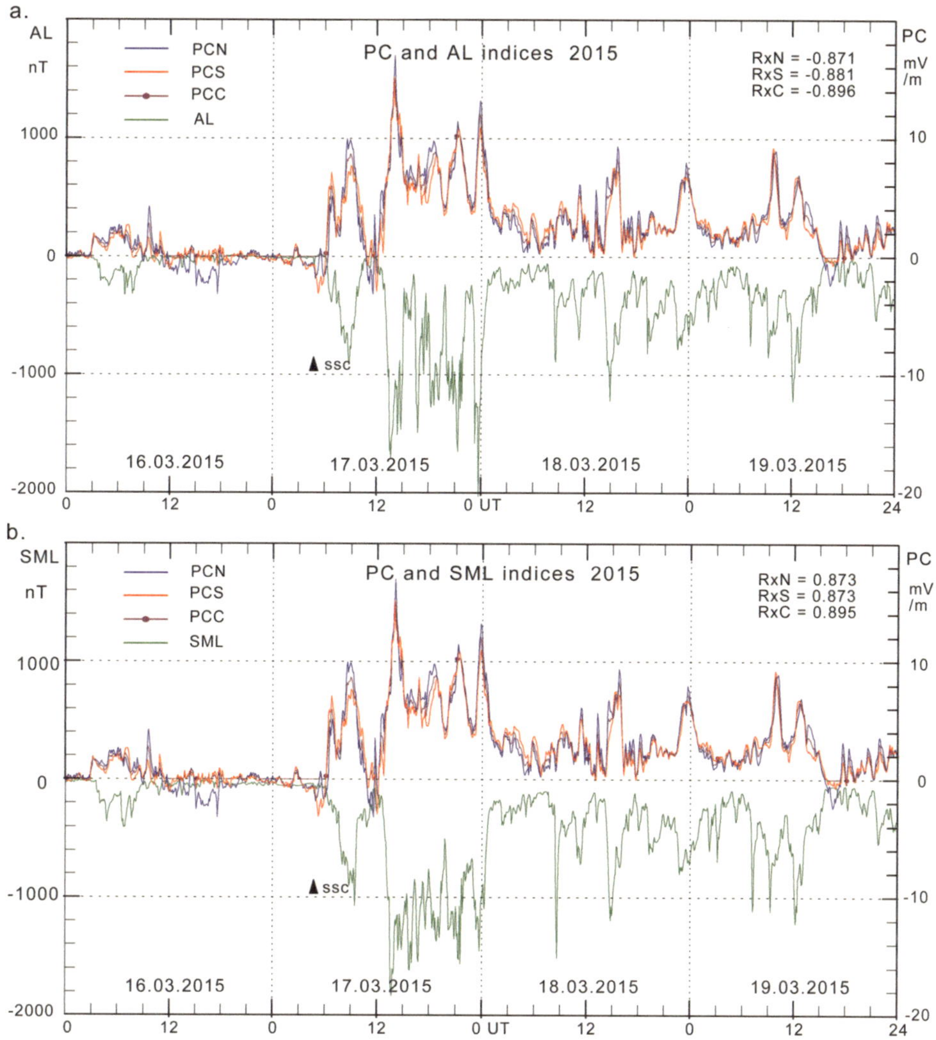

Figure 5.
(a) 5-min samples of PCN (blue), PCS (red), PCC (magenta), and AL (green) indices for 16–19 March 2015. Correlation coefficients are noted at the upper right section. (b) Corresponding display of PCN, PCS, PCC, and SML 5-min index samples for the same storm event.

There are several remarkable features in **Figure 5a** and **b**. The PCN, PCS, and PCC indices are almost identical except during the onset of the magnetic storm indicated by the sudden commencement on 17 March 2015 marked by the pointing (and size) of the black triangle marked SSC. The AL indices in **Figure 5a** are almost (negative) mirror images of the PCC values. The SML indices in **Figure 5b** indicate larger and sharper variations than the AL index values in **Figure 5a** probably due to the extended latitudinal coverage toward subauroral latitudes for the SML indices compared to observatory grid extent for the AL indices. The PCC-based correlation coefficients (RxC) noted in the upper right part of the figures are higher than the corresponding coefficients (RxN, RxS) for PCN or PCS, which speaks (again) for using the PCC over PCN or PCS indices in the examinations of large-scale geomagnetic disturbances.

The examples of statistical processing of a larger amount of auroral electrojet data shifted by 3 min versus corresponding PCC values are shown in **Figure 6a–d**. The correlation between the auroral and polar cap indices and parameters for the guiding regression lines are noted in the plots. The corresponding values based on further data sets are shown in **Figure 7**.

From **Figure 6a–d**, note the larger correlation coefficients for the SML and SMU versus PCC correlations than the corresponding correlations coefficients for the AL and in particular the AU indices versus PCC. The AU indices indicate stagnation for PCC indices beyond 5 mV/m probably because the range of observatories for the auroral electrojet (AE) indices misses the stronger AU-defining eastward electrojet events that move equatorward of the array of standard AE observatories.

Displays and calculations corresponding to **Figure 6** have been made for every year between 1997 and 2019 except 2003 where Vostok (or Dome-C) data were not available for PCS calculations. From these displays, the correlation coefficients and values of the slope and intercepts of the regression have been extracted in order to document their variations with time and solar cycle. The variations in correlation and slopes for SuperMag SML and SMU indices (equivalent of auroral indices, AL and AU) throughout 1997–2019 are displayed in **Figure 7**.

The ratio between amplitudes of 5-min AU, AL, and PCC indices similar to the ratio between SMU, SML, and PCC indices displayed in **Figure 7** have been used to

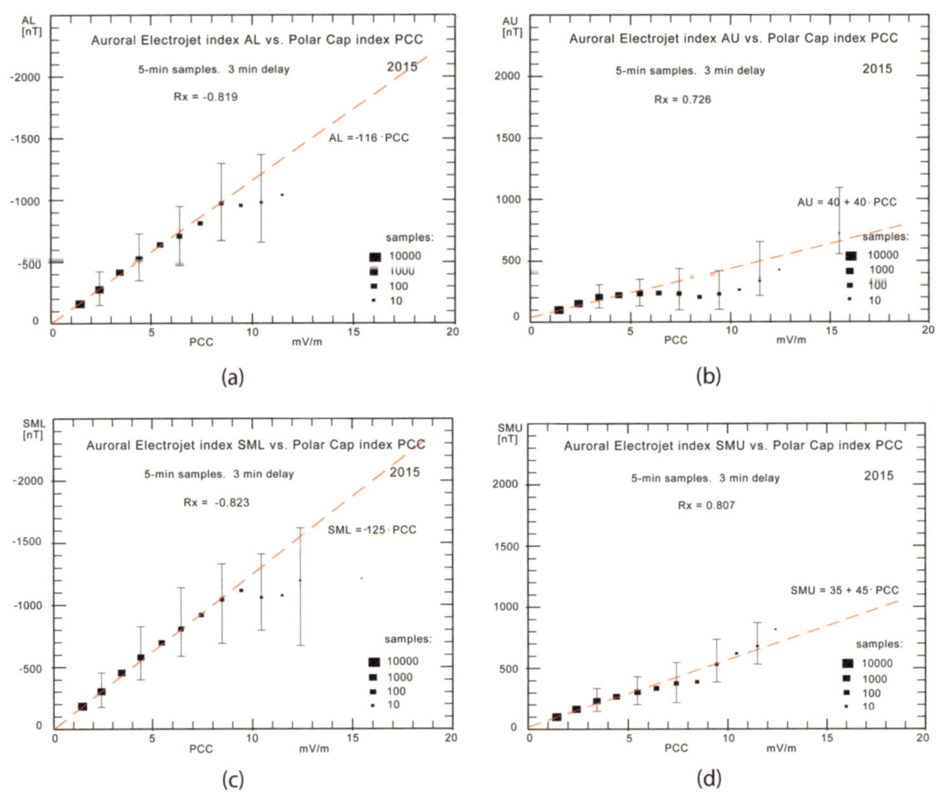

Figure 6.
(a) 5-min AL indices versus PCC. (b) 5-min AU indices versus PCC. (c) 5-min SML indices versus PCC. (d) 5-min SMU indices versus PCC. Correlation and regression results are noted in the diagrams.

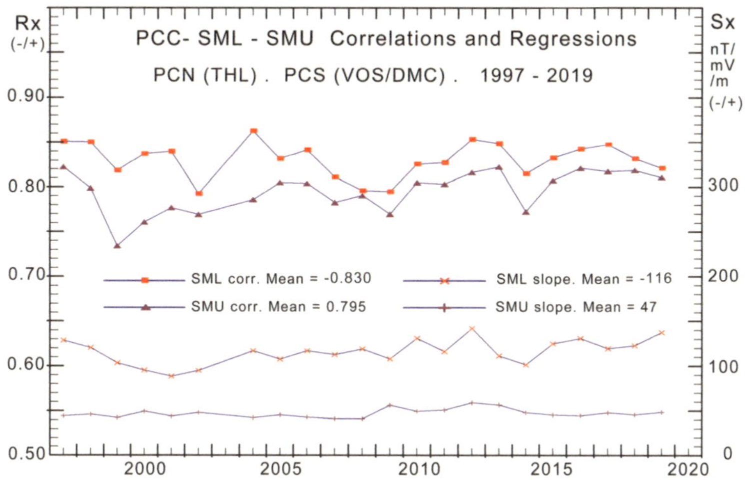

Figure 7.
From top: Yearly average correlations between 5-min PCC and SML (red squares connected by blue lines); yearly average correlations between PCC and SMU indices (magenta triangles); yearly average ratio between 5-min SML and PCC values (red crosses); and yearly average ratio between SMU and PCC indices (magenta crosses).

generate 5-min "equivalent auroral indices," AU_{EQ} and AL_{EQ}, from PCC indices. An example for March 2015 is shown in **Figure 8**.

Figure 8 displays a rather close agreement between the large-scale features of the electrojet, AL, indices and the PCC-based equivalent AL values, while the relations between the electrojet, AU, indices and the PCC-based AU_{EQ} values display less agreement. The use of equivalent auroral indices derived from values of the PCC indices may help to identify disturbances in recognizable appearances in real-time displays used at space weather monitoring. In addition, the conversion enables much closer examinations of the relations between series of the two different parameters, such as their average, absolute, and root-mean-square (RMS) differences, than

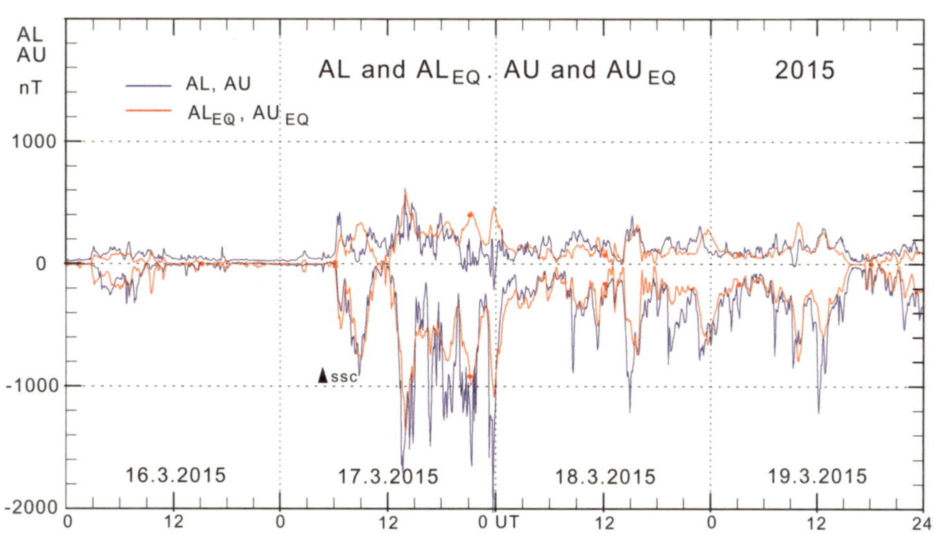

Figure 8.
Upper field: Display of 5-min AU index (blue line) and AU_{EQ} (red line) for 16–19 March 2015. Lower field: Display of AL (blue) and AL_{EQ} (red) indices.

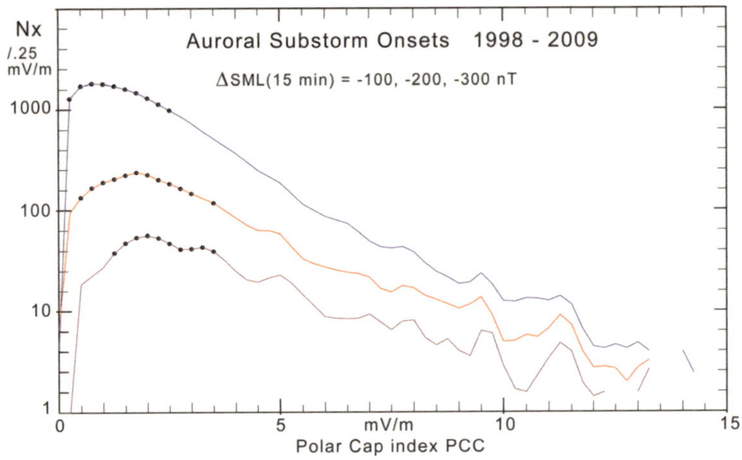

Figure 9.
Number of substorm cases within PCC bins of 0.25 mV/m with SML index changes by more than –100 nT/–200 nT/–300 nT in 15 min versus average PCC index values during the preceding 15 min.

available in just the correlation coefficients. These statistical results may help developing realistic physical models for the relations between solar wind parameters, transpolar plasma convection, and auroral current systems.

The prediction of substorm onset is a particularly intriguing issue. First, relevant substorm onset indications must be defined. There are widely different onset criteria depending on whether the developments in the mid-latitude positive bays [53] or the auroral negative bay [48] are used to identify substorm onsets. In the present work, a change in the 5-min auroral electrojet index, SML, by one of the amounts –100, –200, or –300 nT within 15 min are looked at. The average PCC index values within the preceding 15 min before the onset time are recorded. The number of counts within each bin of 0.25 mV/m in the PCC index is noted. The results from cases in 1998–2009 (ex. 2003) are displayed in **Figure 9**.

The dotted part of the –100 nT curve in **Figure 9** displays the amount of cases above half the top value (1871). This range corresponds to substorm pre-onset levels between PCC ≈0.25 and 2.25 mV/m. This range is broader particularly in the low-level limit than the range of 1.5 ± 0.5 mV/m suggested, among others, in [24] based on the same identification of substorm onset as the criteria used here. The dotted parts of the –200 nT and –300 nT curves indicate, correspondingly, the amount of cases above half the top values of their occurrence patterns.

7. Relations between PC indices and the partial ring current system

The partial (asymmetric) ring current indices, ASY-H and ASY-D, are provided by Kyoto WDC-C2 [51] as 1-min values. Here, the study shall focus on the relations between ASY-H indices and the polar cap indices, PCC. The 1-min samples of both series have been averaged to form 15-min samples. Using a stepwise variable delay between samples of the respective time series, the 15-min index data sets have been subjected to linear correlation analyses assuming that the maximum value of the correlation coefficient provides the most appropriate delay. With this delay imposed on all the pairs of samples of the time series, linear relations between the two parameter sets were found by least squares regression. The average deviation, the average numerical (absolute) deviation, and the RMS standard deviation were calculated from the assumed linear relations.

The reported investigations have considered 4-day intervals of all the major geomagnetic storms with Dst(peak) < –100 nT with onsets from quiet conditions on the first day and occurring during the interval from 1992 to 2018. **Figure 10** displays a scatter plot of 15-min ASY-H index values against PCC values. The number of individual samples for each unit interval in PCC is illustrated by the size of the black squares on the (logarithmic) scale shown in the lower right section of the display. In order to avoid cluttering the display, standard deviation values are shown by the error bars plotted in every other interval only. The 8-min delay noted in **Figure 10** was found to provide least RMS deviation and optimum correlation (Rx = 0.743) for the regression between 15-min samples of the two index series.

A noteworthy feature in the display is the persistent closely linear relation between the average ASY-H and PCC index values up to high disturbance levels. The relation is expressed as

$$\text{ASY}-\text{H} = 10.9 \cdot \text{PCC} + 16\ [\text{nT}]. \tag{9}$$

The number of 15-min samples, correlation coefficients, and results from the linear regression analyses for various PC index versions are summarized in **Table 2** [from [29]]. In addition to presenting PCN, PCS, and the combination PCC [cf. Eq. (8)], the table comprises PCA, which is the plain average of PCN and PCS, and the seasonal selections PSW (local winter) and PCU (summer).

It should be noted that data for the various versions have been selected from the epoch 1992–2018 on the basis of the magnetic storm intervals. Hence, no effort was made to avoid intervals where data for one or the other index version were missing. Note, in particular, the reduced number of PCS samples due to intervals of missing or invalid Vostok data (cf. **Figure 1b**).

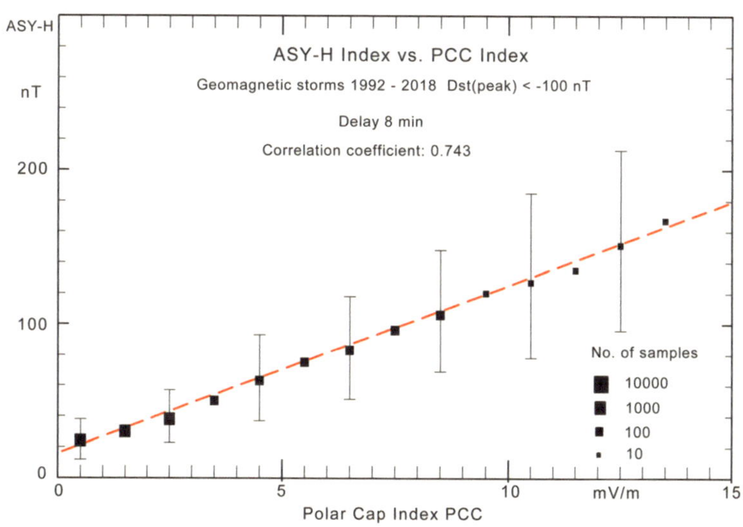

Figure 10.
Scatter plot of ASY-H against PCC index values. The black squares indicate average values and the number of 15-min samples within each unit interval in PCC, while the error bars at every other unit interval indicate standard deviation. The red-dashed line indicates least squares regression slope and offset values for the 15-min data samples [Eq. (9)]. [from [29]].

Version	PCC	PCN	PCS	PCA	PCW	PCU	Unit
Samples	28,803	34,839	28,802	28,880	33,728	29,913	
Correlation	0.743	0.702	0.679	0.716	0.700	0.683	
Mean dev.	–1.7	–0.3	–0.8	–0.5	0.1	–0.6	nT
RMS dev.	23.6	24.5	25.6	24.0	24.0	25.5	nT

Table 2.
Number of samples, correlation coefficients, and regression results for ASY-H/PC relations [from [29]].

While the correlation coefficient for 15-min samples of the PCC-ASY-H relation in **Figure 10** is Rx = 0.743, then the correlation between the ASY-H and further PC index versions is all close to correlation coefficient values of only around Rx = 0.70. The problem resides, in particular, with the negative PC index values since the ASY-H values are dominated by positive index values.

The linear relation between ASY-H and PCC could be used to generate an "equivalent ASY-H index," ASY-H_{EQ}, based on rescaling the PCC indices according to Eq. (9). This would provide basis for close comparisons of the two index series by enabling the calculations of parameters such as mean, absolute, and RMS deviations at various conditions. The calculation of equivalent values might, furthermore, generate a display of ASY-H_{EQ} looking much like displays of the real ASY-H indices, which could be useful for monitoring the asymmetric ring current developments in real-time applications where the PC indices are available online in their real-time version. The examples of real (published) ASY-H indices and PCC-based ASY-H_{EQ} indices are displayed in **Figure 11a** and **b** for the magnetic storm events of 16–19 March 2015 and 22–25 June 2015, respectively.

Comparing ASY-H (magenta line with dots) with ASY-H_{EQ} (red line) in the upper fields of **Figure 11a** and **b** indicates a moderate degree of agreement, which is best at the onset phase of the displayed magnetic storms starting on 17 March 2015 and 22 June 2015, respectively. The direct conversion of PCC indices to equivalent SYM-H values shown in the lower fields is not working well (see Section 8).

8. Relations between PC indices and the symmetrical ring currents

The intensity of symmetrical ring currents can be monitored by the hourly ring current index, Dst, and more detailed by the 1-min SYM-H indices provided by Kyoto WDC-C2 [51, 54]. The relations between series of polar cap indices and SYM-H (or Dst) indices using a variable delay (PC indices leading) have failed to generate maximum correlation at time shifts by up to 4 h [29]. Instead, the approach suggested in [27, 28] is applied here. Thus, the PCC index is used in a source function for the gradient in the Dst index rather than in correlations with its actual values.

The Dst index [55] is considered to represent the amount of energy stored in the ring current by the Dessler–Parker–Sckopke relation [56, 57]. Following Burton et al. [58], the rate of change in the Dst* index with time could be written as

$$dDst^*/dt\,[nT/h] = Q[nT/h] - Dst^*[nT]/\tau\,[h]. \tag{10}$$

Here, Dst* is the Dst index corrected for contributions from magnetopause currents (MPC). The quantity Q (in nT/h) is the source term, while the last term in

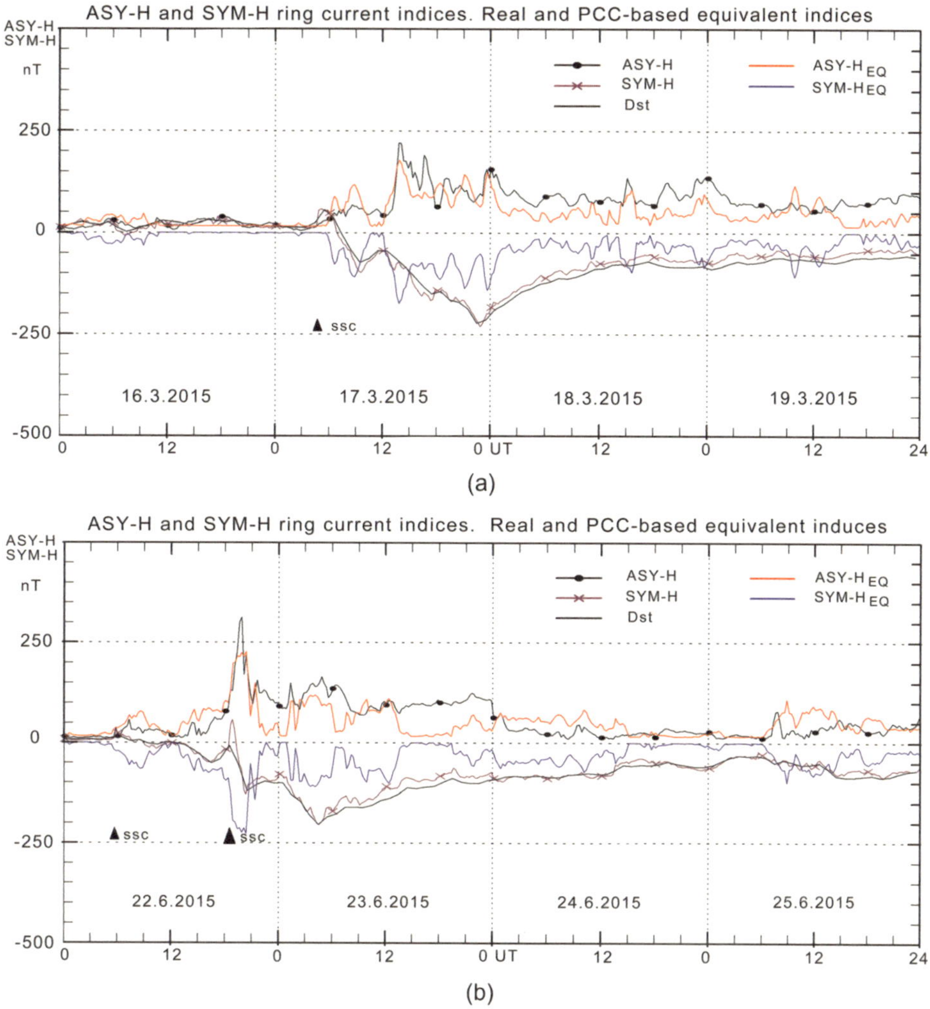

Figure 11.
Magnetic storms (a) 16–19 March 2015 and (b) 22–25 June 2015. The upper part of the fields display real (published) ASY-H indices (black line with dots) and equivalent ASY-H$_{EQ}$ values (red) converted from PCC index values by scaling [Eq. (9)]. The lower parts display real (published) SYM-H (magenta line with crosses) and Dst indices (black line), and equivalent SYM-H$_{EQ}$ index values (blue) converted from PCC indices by scaling. The triangular symbols mark events of storm sudden commencements (SSC).

Eq. (10) is the ring current loss function controlled by the decay time constant, τ, [59]. For the small actual MPC corrections, the Dst-dependent statistical values provided in [60] are used here. For further details, see [29]. Now, the relation in Eq. (10) has only terms relating to the source function Q and may provide equivalent Dst index values by integration from a known state, once the source term is defined.

In [58], the source term Q was related to the Y_{GSM} component of the solar wind electric field. In the analyses by Stauning et al. [27] and Stauning [28], the relations of Q to the polar cap indices were examined for a number of storm event cases during the intervals 1995–2002 and 1995–2005, respectively. These analyses were extended in [29] to comprise selected large storm events with Dst(peak) < –100 nT and storm onset on the first day throughout 1992–2018 in order to improve the statistical basis. The temporal change at time t = T in the hourly Dst* index were derived from the hourly values at t = T – 1 and t = T + 1 [h] by the simple differential term:

$$dDst^*/dt(T\,h) = (Dst*(T+1\,h) - Dst^*(T-1\,h))/2. \quad (11)$$

In **Figure 12**, hourly values of dDst*/dt derived from archived data using Eq. (11) and corrected for decay (Eq. (10); [59]) have been plotted against the related PCC index values. Average values within each unit of PCC are displayed by the black squares with sizes corresponding to the number of hourly samples according to the lower right scale, while standard deviation is marked by error bars (in every other bin).

The scatter plot in **Figure 12** presents variations in the Dst* source function, Q_{OBS}, with PCC using a variable time shift to obtain the best correlation. The relation between the best fit source function, Q_{OBS}, and the source parameter values, PCC, is then expressed in a linear function. From the selected data set (98 storm periods within the interval from 1992 to 2018), the regression on the total amount of hourly samples provides the slope and offset values indicated by the red-dashed line in **Figure 12** and reported as follows:

$$Q_{OBS}\,[nT/h] = -4.1\,[(nT/h)/(mV/m)] \cdot PCC\,[mV/m] - 2.2\,[nT/h]. \quad (12)$$

This result is close to the corresponding source function (Q = –4.6·PCC – 1.2) defined in Stauning [28] from a smaller amount of data (storm events 1995–2005).

With continuous time series of the PCC-based source values, and specifications of the relational constants and initial Dst values, it is now possible, at least in principle, to integrate Eq. (10) to derive the values of an "equivalent Dst index", Dst_{EQ}, throughout any interval of time. The reported work [29] has brought the analysis of the relations between Dst and the polar cap index, PCC, important steps forward compared to [28] by including a close examination of the decay time constants (τ = 5.8 h and τ = 8.2 h) in [59] and their turning level (DstXlevel = –55 nT), and other

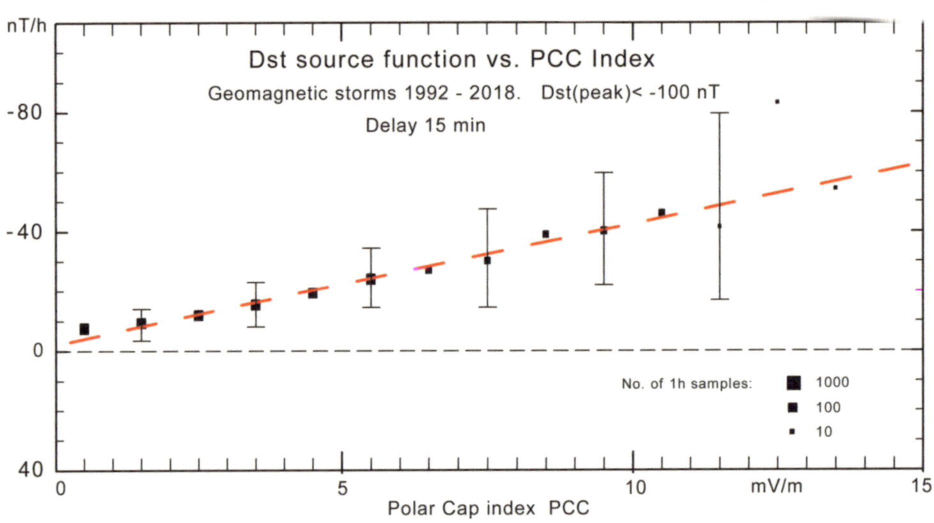

Figure 12.
Scatter plot of d(Dst)/dt corrected for decay versus polar cap PCC index. The black squares represent bin-average values and no of hourly samples, while the error bars in every other bin represent standard deviations (from [29]).*

parameters of importance for the relations between Dst and its possible source functions, primarily the PCC index. A further parameter introduced here is the optimum delay between samples of the PCC time series and the Dst values. For these cases, the PCC-based index values lead by a few (≈45) min.

In addition to the decay time constants [59], the examination included the impact from the saturation [61] of the PC indices at high levels of the merging electric field, E_M, as seen in all PC index series [18].

In a crude approximation for the parameter iteration process, an "effective PCC index" (PCCeff) were set equal to the E_M values up to a turning level (PCClim) at around 5 mV/m and then forced to deviate by adding a linearly varying term with slope (S) less than unity. The approximation is defined by the two-step linear relation as follows:

$$\text{PCCeff} = \text{PCC for PCC} < \text{PCClim} \tag{13}$$

and

$$\text{PCCeff} = \text{PCC} + \text{Seff} \cdot (\text{PCC–PCClim}) \text{ for PCC} > \text{PCClim} \tag{14}$$

where Seff = (1/S – 1) is less than unity.

The above-mentioned set of 98 magnetic storms with peak Dst below –100 nT occurring throughout the epoch from 1992 to 2018 where PCS indices are available (with some gaps) was used as a testbed to explore the effects of parameter adjustments. For the calculation of PCN indices, Qaanaaq (THL) data have been almost continuously available since 1975. Dome-C magnetic data have been substituted for missing or unreliable Vostok data (cf. **Figure 1b**) for PCS calculations, among others, throughout 2012 and 2013. For each storm event, a sequence of 4 days is considered at a time with the storm beginning from quiet conditions on the first day. Starting from the initial decay time values defined in [59], the source values defined in Eq. (12), and the PCC values modified according Eqs. (13) and (14), the parameters were changed in small successive step searching for maximum correlation and minimum deviations between real and equivalent Dst values.

The examples of observation-based and equivalent Dst values are displayed in **Figure 13a** and **b**. For these cases, the integration of the source term has been started at the real Dst value and then allowed to proceed independently throughout the 4 days in each set.

The examples in **Figure 13a** and **b** were based on using PCC indices in the source function for possible applications of the technique at space weather monitoring. They represent cases (a) of high correlation (Rx = 0.957) and (b) moderate correlation (Rx = 0.762) compared to the average correlation level (Rx = 0.810) for the selection of storm events. A specific feature of **Figure 13b** is the effects of the strong storm sudden commencement (SSC) at 15 UT on 22 June 2015. The occurrences of SSC events counteract the ring current effects on low-latitude magnetic observations, thereby preventing the reported Dst index values from reaching the full (negative) peak values corresponding to the actual ring current intensities displayed by the PCC-based Dst_{EQ} index series, which is less sensitive to SSC effects.

Figure 13a and **b** indicate very good and fair agreement, respectively, between the real and the equivalent Dst values. Generally, the agreement is best for moderate storms. Going from the moderate to the strong storm cases gives sometimes less agreement between real Dst values and equivalent PCC-based Dst_{EQ} values, possibly related to saturation effects not compensated for by the PCC modifications defined in

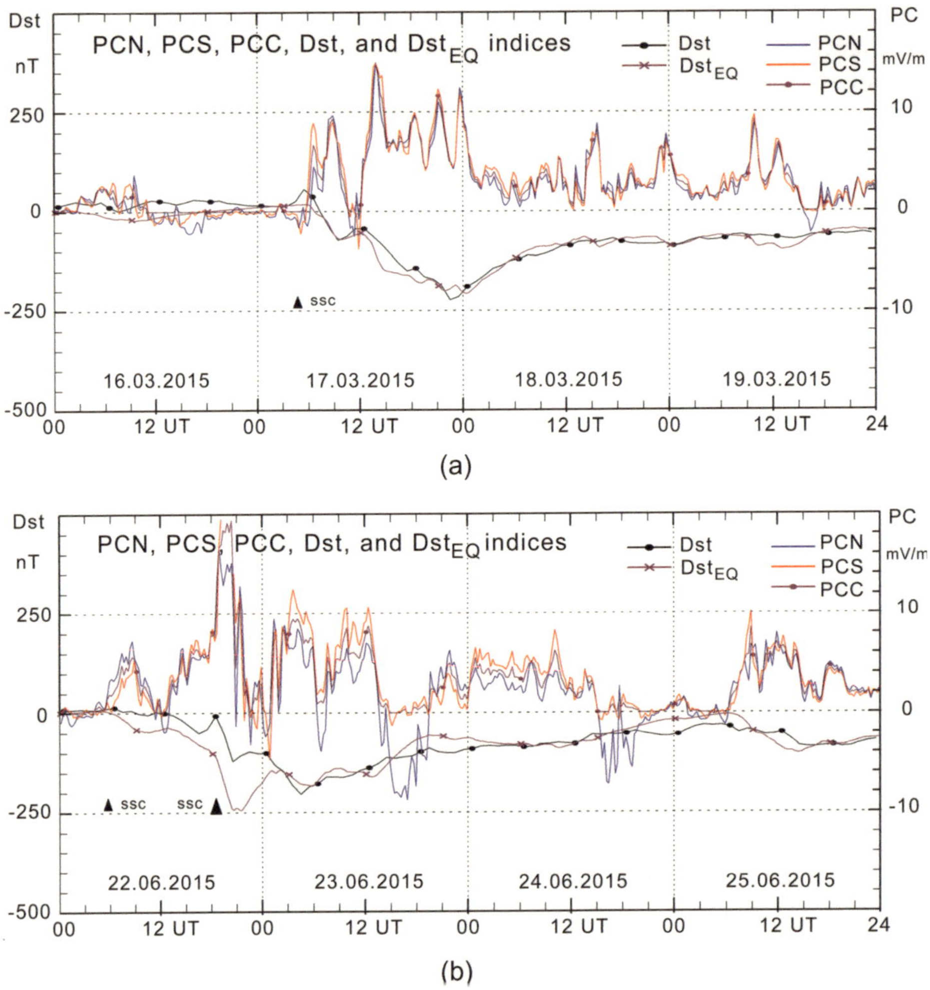

Figure 13.
(a, b) Examples of published (real) Dst (black line, dots) and equivalent Dst (magenta) values calculated from the PCC-based source function. Values of PCC (magenta), PCN (blue), and PCS (red) are displayed in the upper fields on the right scale [from [29]].

Eqs. (13) and (14). For the very weak cases, the uncertain influence from magnetopause currents (MPC), although small, may have relatively large effects, which reduce the agreement between the real and the equivalent Dst indices.

With the understanding of the effects of adjustments of the various parameters gained from the test bed exercises, the full range of available data has been used to integrate the PCC-based source function throughout the epoch from 1992 to 2018 to derive equivalent Dst_{EQ} values without attachment at all to the published (real) Dst values. In cases where either PCN or (Vostok or Dome-C-based) PCS values were unavailable, the available hemispherical PC indices were used for PCC index calculations with degraded accuracy.

In the first step, the timing parameters were adjusted to provide the overall best correlation and least deviations related to values of decay time constants for the "fast" decay at high disturbance levels and "slow" decay at low disturbance levels as well as

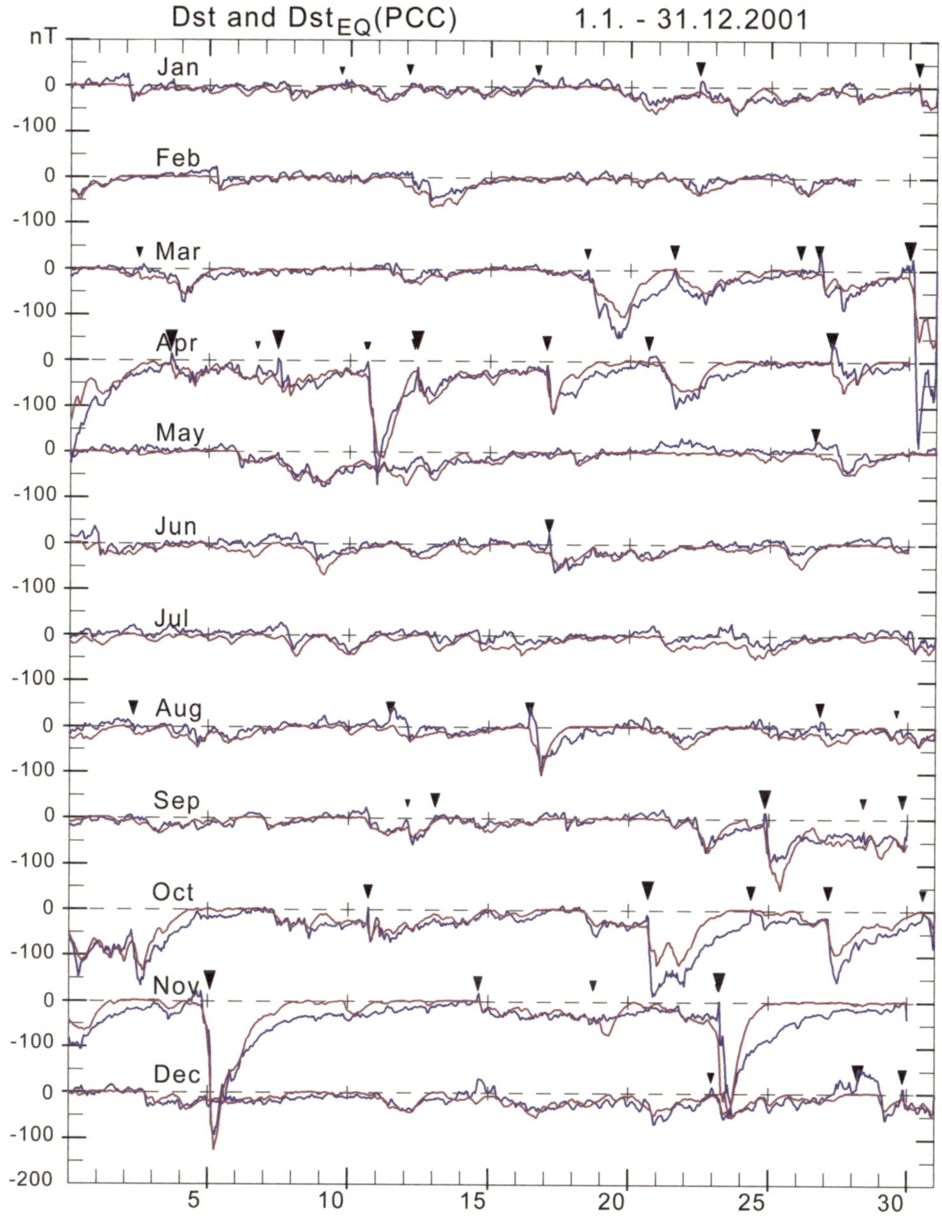

Figure 14.
Observed Dst values (blue line) and calculated Dst_{EQ} values (magenta) for 2001. Storm sudden commencement (SSC) events are displayed by the downward pointing black triangles to indicate onset times and sized to indicate their amplitudes [from [29]].

Dst values at the cross-over (DstXlevel). In a second step, the PCC high-level modifications suggested by Eqs. (13) and (14) were used to provide the best possible agreement between peak values of Dst_{EQ} and Dst keeping the other parameters near their initial values. The iterations gave slightly different parameter values depending on the choice of quality parameters considered in the process. Thus, the final parameter values are not unique but represent compromises. For the control of the methods and calculations, the derived Dst_{EQ} values have been displayed in plots along with the real

Symbol	Optimal	Unit
Fast decay, τ_1	5.5	h
Slow decay, τ_2	7.0	h
DstX level	–52	nT
Dst gradient	–4.5	(nT/h)/(mV/m)
PCClim	5.0	mV/m
PCCslope, S_{eff}	0.60	–
Delay Dst_{EQ} – Dst	45	min

Table 3.
Parameters for Dst_{EQ} calculations.

Result term	Optimal	Unit
Mean Dst	–13.08	nT
Mean Dst_{EQ}	–13.09	nT
Mean diff.	–0.01	nT
Abs. diff.	8.88	nT
RMS diff.	12.30	nT
Correlation	0.856	–

Table 4.
Results from Dst_{EQ} calculations.

Dst values throughout the entire epoch. Interim examples for the stormy year 2001 are displayed in **Figure 14**.

Figure 14 displays close, although not perfect, match between the real Dst values (blue line) based on observed near-equatorial magnetic variations and the equivalent Dst_{EQ} values (magenta line) calculated by the integration of the PCC-based source function [Eq. (10)] using the parameters estimated from the lengthy correlations throughout 1992–2018 including high-level modifications of the PCC indices. The integration was performed in steps of 5 min starting from Dst_{EQ} = 0 on 1 January 1992 using parameters listed in **Table 3**. A summary of results is listed in **Table 4**.

In **Tables 3** and **4**, the "Optimal" columns refer to parameters and results derived from the total 1992–2018 integration sequence [from [29]].

9. Real-time applications

When the relations discussed in the preceding sections are used in real-time applications, such as the monitoring and forecasts of space weather conditions, then it is necessary to derive the PC index parameters in real-time versions based on polar magnetic data assumed available in real time either directly from an observatory or from Internet links.

The scaling parameters are the same being independent on whether used for post-event or real-time calculation of PC indices. The secularly varying baseline values are predictable to a high degree of accuracy. Thus, the problem resides with the QDC-level calculations.

As mentioned in Section 4, the solar rotation weighted (SRW) QDC method is very well suited for real-time applications. While maintaining the weight factor definitions, the calculation scheme is not changed by the transition from using quiet samples from the past 40 days up to the present hour neglecting post-event samples in real-time PC index calculations to using the full range of data from ±40 days for post-event index calculations.

The differences between the post-event and real-time PC indices were examined in [62]. **Figure 15a** displays in the upper two panels an example for 2015 of hourly PCS index values derived by using post-event (PE) and (simulated) real-time (SRT) methods, respectively. The bottom panel displays the differences. These differences are largest in the local summer months.

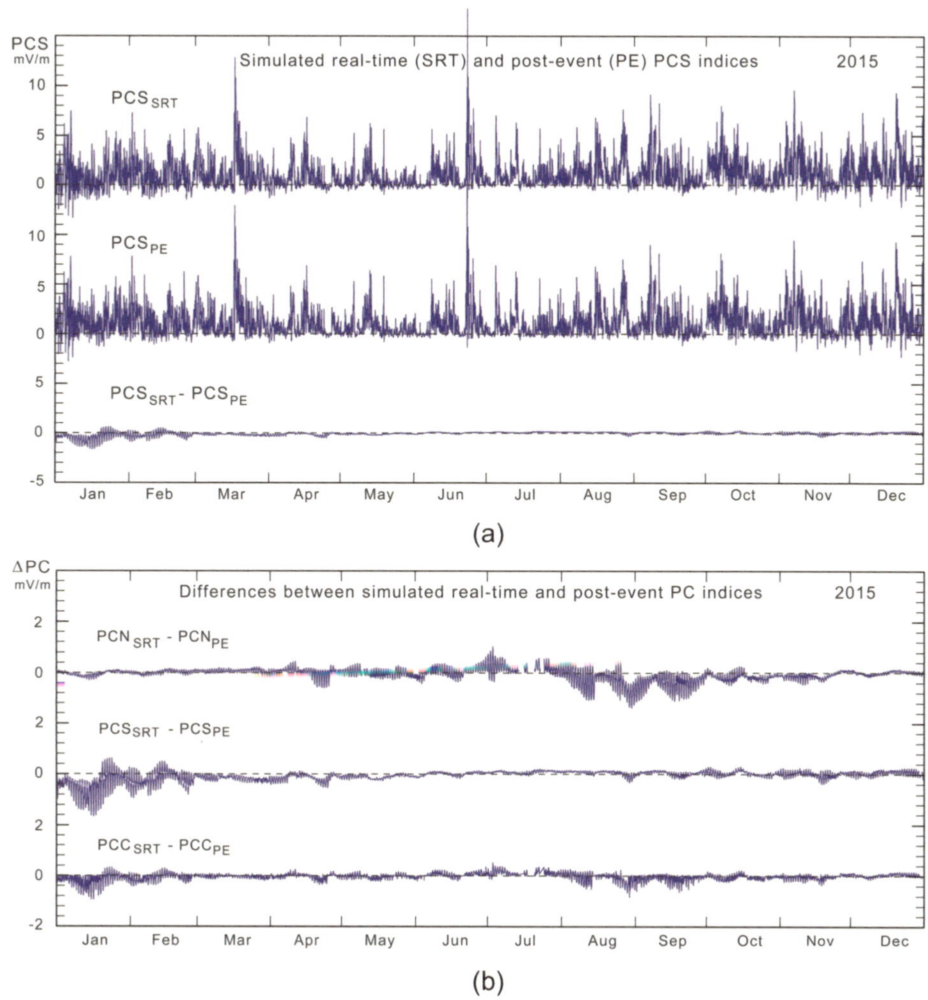

Figure 15.
(a) Simulated real-time (SRT) (upper part) and post event (PE) PCS (middle part) index values based on Vostok magnetic data using common calibration parameters and SRT or PE versions of QDC values, respectively. The lower part of the field displays their differences. (b) Differences between SRT-based QDCs and PE-based QDCs for PCN, PCS, and PCC indices [from [62]].

The differences for PCS have been transferred to the middle field of **Figure 15b** on a more sensitive scale. The corresponding calculations of differences were made for the PCN indices presented in the upper field of **Figure 15b** and for the PCC indices presented in the bottom field of **Figure 15b**. It is seen that the relatively small and seasonally dependent differences between the post-event and the real-time PCN and PCS versions are further mitigated when derived for the combined PCC index.

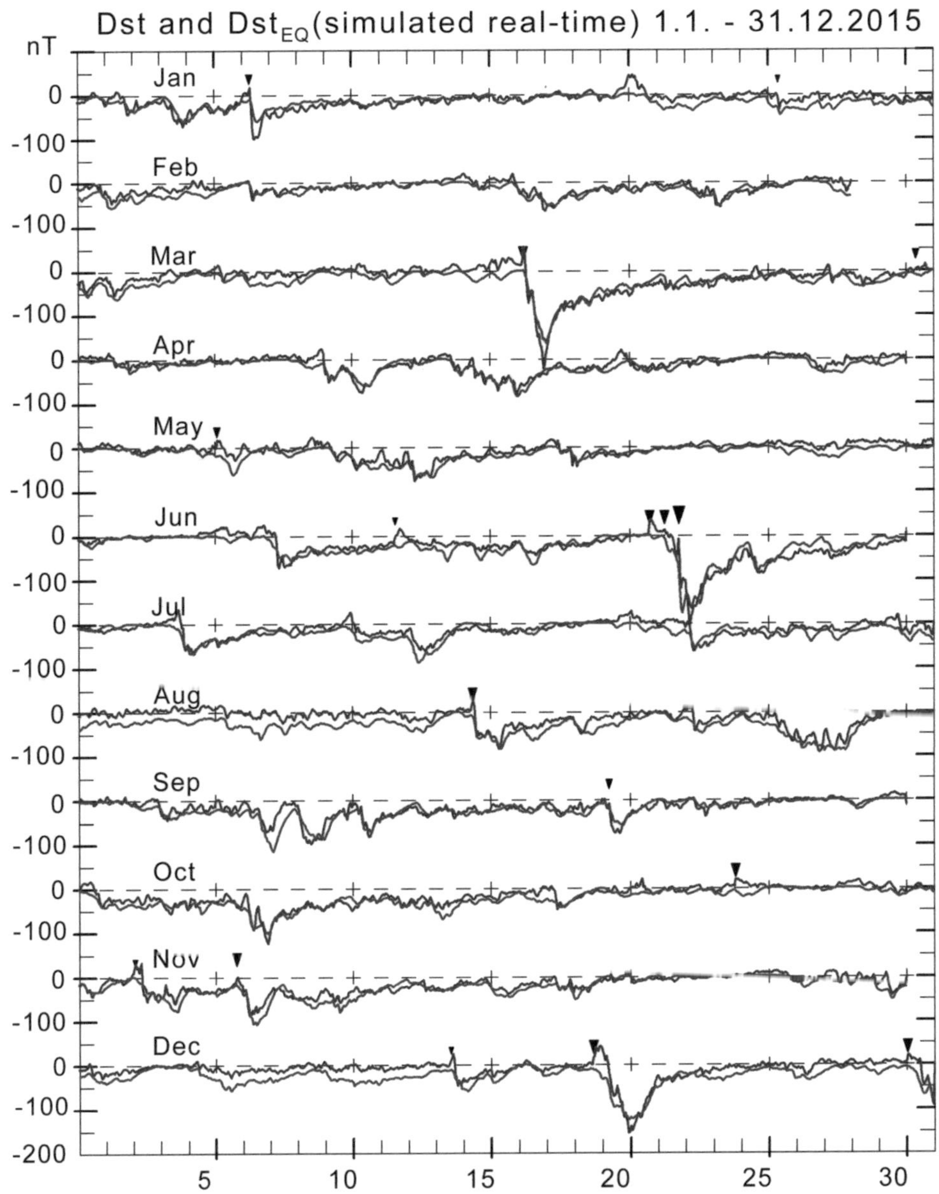

Figure 16.
Real (published) Dst values (blue line) and calculated PCC-based Dst_{EQ} values (magenta) in simulated real time for 2015. Storm sudden commencement (SSC) events are displayed by the downward pointing black triangles to indicate onset times and sized to indicate their amplitudes [from [62]].

Thus, the use of real-time PCC indices will not change in any significant amount any of the relations between the polar cap indices and the magnetospheric current systems discussed in the present contribution mostly based on post-event analyses.

Using QDC values calculated in the real-time version, that is, from past days only with respect to current time, enables the calculation of PCN and PCS and, thus, PCC indices in simulated real-time versions. These versions have been used in the lengthy integration from 2009 to 2019 to calculate PCC-based equivalent Dst indices. An interim result for 2015 is displayed in **Figure 16** in the format of **Figure 14**.

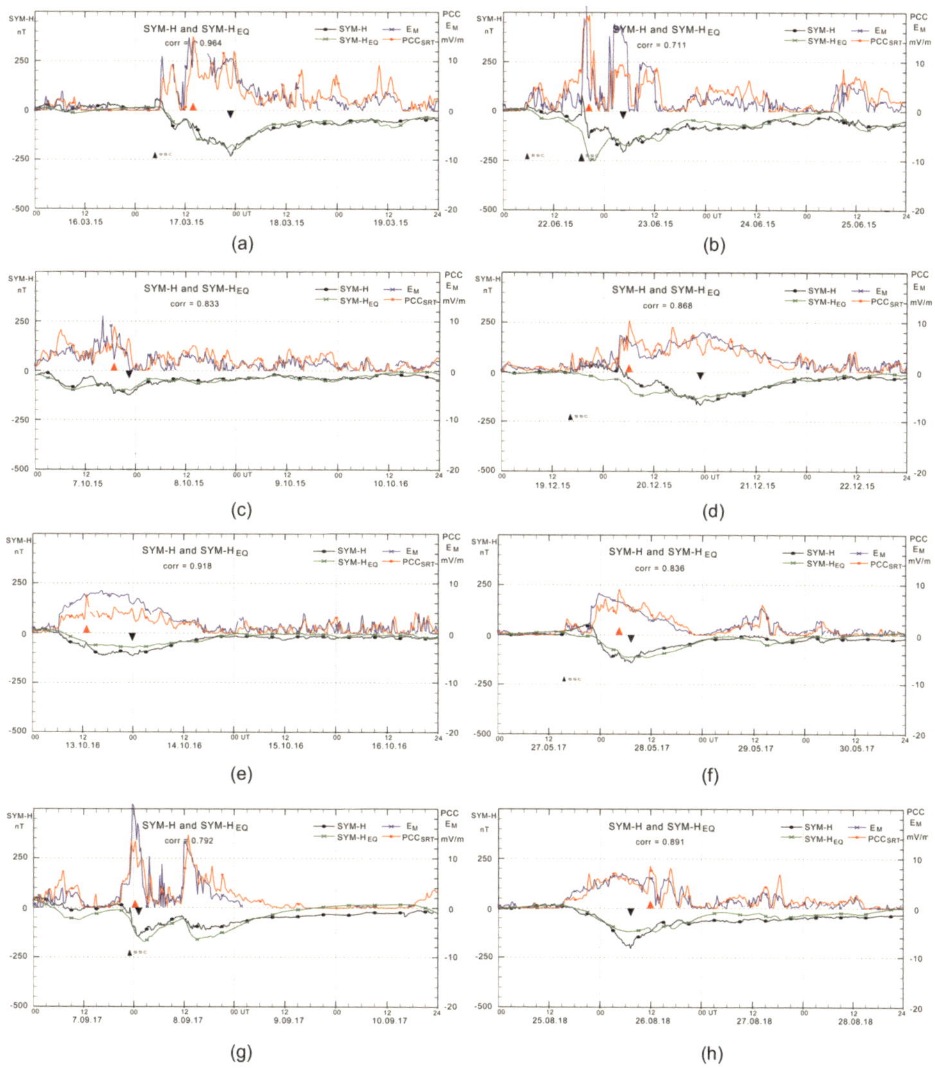

Figure 17.
(a–d) Examples of storm events. Low-latitude observations-based (real) SYM-H indices (black line with dots) and equivalent SYM-H_{EQ} index values (green, crosses) calculated from the source function [Eq. (10)] using the PCC_{SRT} values (red) shown in the upper fields along with merging electric field values, E_M, (blue), displayed on the right scale. Upward pointing red triangles at the mid-line indicate PCC maxima. Downward black triangles indicate SYM-H minima. The size and pointing of the small black triangles labeled SSC indicate intensity and onset time of storm sudden commencements [after [62]].

The match between the real and the simulated Dst values in the simulated real-time example shown in **Figure 16** is as good as seen in the corresponding post-event example displayed in **Figure 14**.

The use of simulated real-time calculations of SYM-H indices is illustrated in **Figure 17a–d** for a sequence of storm cases observed between March 2015 and June 2018. The diagrams present, in the lower fields, the values of SYM-H (black line with dots) and SYM-H_{EQ} (green line). The upper fields of the diagrams in **Figure 17** present the PCC indices and, in addition, the values of the merging electric fields, E_M calculated from OMNIweb solar wind parameters (http://omniweb.gsfc.nasa.gov).

It the examples in **Figure 17**, there are several noteworthy features such as the close agreement between the real SYM-H values and equivalent SYM-H_{EQ} values derived from PCC_{SRT} indices calculated in simulated real-time using past values only in the calculations of QDC reference levels. The diagram presents the displays that would be seen in real-time space weather monitoring with access to polar magnetic data, space data to calculate E_M, and real SYM-H values in real time. The agreement between PCC and E_M values underlines their close correlation, as displayed in **Figures 3** and **4**. If space data for calculations of E_M and real SYM-H values are not available, then PCC and the PCC-based equivalent SYM-H values might provide worthwhile substitutes at space weather monitoring.

Note, in passing, the scatter in the relative occurrences of maxima in PCC and minima in SYM-H as depicted by the positions of the upward pointing red triangles and the downward pointing black triangles at the midline. This scatter is in strong contrast to the regularity in timing claimed by Troshichev and Sormakov [26] and included in the report ISO-TR/23989 from the International Organization for Standardization (ISO) [35].

In the publication and in the ISO report, they state that maximum depression (minimum SYM-H) would occur "*with typical delay time of* $\Delta T = 1.5 \pm 0.5$ *h*" after the peak in the PC index. This statement is obviously incorrect for all but one of the eight cases presented in **Figure 16** (six cases of ΔT = 4–14 h, 1 case of ΔT = –6 h). The invalid statements from Troshichev and Sormakov [26] included in the ISO [35] report could be seriously misleading at space weather monitoring when trusting that the ISO represents the supreme authority in space environment.

10. Discussions

10.1 PC index concept

Using the Kan and Lee [21] merging electric field (energy coupling function) for scaling of the index calibration parameters is a fundamental premise for the development of the present PC index concept [10]. However, it is quite possible that the reference energy coupling function could be improved. Further coupling functions, like those suggested by Lyatsky and Khazanov [63], Tenfjord and Østgaard [64], or McPherron et al. [65], might provide more precise relations between solar wind parameters and geomagnetic disturbances but at the cost of simplicity. The task of comparing the performances of different energy coupling functions awaits a dedicated initiative.

However, the results for the relations between polar cap indices and indices for further major magnetospheric current systems reported here strongly favor the present PC index concept expressed in Sections 2 and 3.

10.2 Correlation techniques

For the works reported in present manuscript, all the correlations were made by using a linear product-moment formula. Most regression calculations were made by applying linear least squares regression using the basic sample types considered most useful for the purpose. For the correlation and regression calculations for PC indices against solar wind parameters and global magnetic disturbance indices, 1-min data samples are available. However, the faster variations may not be transferred in complete details between the solar wind and the polar ionosphere or between the polar cap and the ring current regimes. Thus, for PC-E_M, PC-ASY-H, and PC-SYM-H relations, 15-min samples were used, while 5-min samples were used for the relations between the PC indices and the auroral electrojet intensities and for the onset conditions defined through the AL, AU, SML, and SMU indices. The PC index data were first converted from 1-min to 5-min averages by removing the max and min values for spike suppression. Next, 5-min or 15-min averages were formed from the remaining samples assuming that similar spike suppression measures have been applied to the other parameters by the index suppliers.

10.3 Forward versus reverse convection conditions

The present work discusses the statement in Resolution no. 3 [66], that International Association of Geomagnetism and Aeronomy (IAGA) is "*considering that the Polar Cap (PC) index constitutes a quantitative estimate of geomagnetic activity at polar latitudes and serves as a proxy for energy that enters into the magnetosphere during solar wind-magnetosphere coupling.*"

The first part of the statement is evident, while the second part is based on the close relation between the solar wind merging electric field parameter (E_M) and PC index values as well as the association between PC index levels and various energy dissipation processes, such as auroral activity and building ring currents [23–26, 35, 67–69]. However, the occurrences of negative PC index cases are usually ignored and left out in these associations, and the question on interpretation of negative index values remains open.

It is a fundamental issue for the polar cap index concept that the antisunward transpolar forward convection mode (DP2) at southward IMF is fundamentally different from the reverse convection mode (DP3) associated with northward IMF conditions. The differences appear in the correlation and regression relations between the non-negative values of the merging electric field [cf. Eq. (3)] and the projected magnetic variations that could have positive as well as negative (or even strongly negative) values. In the forward convection cases (positive PC indices), the disturbance level rises with increasing values of the merging electric field that controls the input of solar wind energy into the magnetosphere. In these cases, the PC indices track the merging electric field values. However, as the IMF turns northward (positive B_Z), the transpolar convection may turn sunward (reverse), whereby the PC indices may reach large negative values that could not possibly keep any proportionality with the decreasing but still positive merging electric field values. The examination of the relations in [29] has shown that reverse convection intensities amount to around 3% of the forward convection intensities for Vostok (PCS) and 10% for Qaanaaq (PCN) on the average. However, at daytime in the summer season, the relative amounts may rise to 6% for Vostok and up to 25% for Qaanaaq.

These differences between DP2 and DP3 cases were not implemented in the version [42] submitted jointly from AARI and DTU Space for endorsement by IAGA and granted by its Resolution no. 3 [66]. When reverse convection cases are included, then the adverse effects on the calculations of scaling parameters cause, among others, uneven daily and seasonal relations between PC index values and values of E_M. The relative frequency of reverse convection cases is highest in the daytime hours of the summer season causing the adverse reducing effects on the PC-E_M correlation demonstrated in **Figures 3** and **4**.

The recognition of the differences between forward and reverse convection modes brought forward the PCC index concept Stauning [15] used extensively in the present contribution. These differences have also prompted the development of new calculation schemes for derivation of PCN and PCS scaling parameters (φ, α, and β), as reported in [14, 17, 70]. In the selection of samples from the epoch used for the calculation of PC index scaling parameters, cases of strong northward IMF (NBZ) conditions were omitted as far as possible.

In addition to differences in the calculation of PC index scaling parameters, the definition of the reference level, from which the magnetic disturbance values involved in calculations of the PC indices are measured (cf. Section 4), also differs between the IAGA-recommended PC index derivation methods [42, 71–74] and the SRW method [16] applied to derive reference levels ("QDC"s) for calculations of the indices considered here. These differences are elaborated in [20], where the problems are discussed. The main obstacle is the use of a solar wind sector term introduced by Janzhura and Troshichev [25] and used in the index calculations reported in [42] (see [20, 70, 75, 76]).

The good results from using PCC indices in auroral and ring current mapping reported here further support the concept of PCC indices being the optimum choice for estimates of solar wind energy input in post-event as well as in real-time applications. With the possibility of using data from Dome-C for useful PCS indices and Alert or Resolute Bay data for PCN indices [18, 19], the availability of useful PCC index series is greatly improved. Since 2009 and up to present (2022), there is hardly any interval without useful PCC index values. The present contribution has used data from all these sources with common unified derivation schemes. Comprehensive data and index quality control measures have been implemented throughout, among others, by comparing PC index values derived from different sources, thereby avoiding the invalid indices haunting the IAGA-endorsed PC index series [19, 20].

10.4 PC indices and the 1-min ring current indices

Building the ring currents flowing near equator at distances of 4–6 Earth radii (R_E) is usually considered a feature related to the amount of energy supplied from the solar wind to the magnetosphere [56–60]. For the asymmetrical ring current index, ASY-H, **Figure 10** displays a close relation between the average values of the polar cap PCC indices and the ASY-H indices all the way from near zero during quiet conditions to high values of both indices representing magnetic storm cases.

The direct relations between the SYM-H index and the PCC indices are inconclusive apart from indicating increased SYM-H values with increased PCC values (to be expected). Direct correlations between PC indices and SYM-H or Dst values beyond this trivial relation are not meaningful. This view is actually supported by the analysis of the relations between SYM-H and PC (here average of PCN and PCS) indices presented in [26]. Their **Figure 1** displays the level of correlation between SYM-H and PC indices with varying degree of smoothing of both parameters. The correlation coefficient increases steadily from 0.590 at 15-min samples through 0.625 at hourly average samples to reach 0.657 (which may not be the maximum) at 120-min sample averaging. Thus, their correlation coefficient values, e.g., Rx = 0.625 at hourly average samples, agree well with the estimates in [29] of Rx = 0.623 (which is not maximum) for 15-min samples and 1-h shift and support the conclusion that the low-correlation coefficients indicate poor correspondence between simultaneous PC and Dst (or SYM-H) index values.

Guidance by the calculations presented by Burton et al. [58] has enabled the calculations of equivalent Dst index values based on using PC indices in the gradient source function and in the first step using the timing parameters from Feldstein et al. [59]. The iteration of control parameters to reach an optimal result for the equivalent Dst index depends on the quality parameter considered, whether being the coefficients of correlation with the published (real) Dst values or the mean and rms differences with respect to the equivalent (PC-based) Dst (or SYM-H) values. The relations also depend on the selection of samples considered. The Dst_{EQ} for moderate events would match the real Dst values closer than seen in the very strong or rather weak events. Thus, there is no unique set of "correct" control parameters. The set of values presented in the "optimum" column of **Table 3** is considered the best compromise.

10.5 The PC index as indicator of solar wind energy input

In IAGA Resolution #3 [66] as well as in many reported investigations, the solar wind merging electric field E_M (often named E_{KL}) derived from solar wind parameters at the front of the magnetosphere is considered to control the amount of solar wind energy that enters the magnetosphere. Thus, disregarding negative values, the polar cap indices, since they are scaled with respect to E_M, have been considered to represent the input of solar wind energy to the magnetosphere to power various geomagnetic disturbances such as polar magnetic variations, auroral activities, substorms, upper atmosphere heating, and the building of ring currents [10, 11, 23–25, 68, 77]. Such investigations were previously based on separate PCN or PCS indices, or on their plain averages, or on the summer/winter hemisphere PC index selections. Thus, the improved correlation with E_M resulting from using the PCC indices as reported here might improve results from such investigations.

Some of the relations between PC indices and geomagnetic disturbances could be described to a fair approximation by linear relations possibly with timing shifts to account for propagation delays or inertia effects. However, the gradual building of the symmetrical ring currents to represent energy input from the solar wind to the magnetosphere supports the concept of using the polar cap indices in a gradient source function rather than comparing the PC indices directly to the actual ring current indices. The development of the Dst ring current indices, in particular their

Correlation	PCC	PCN	PCS	PCA[1]	PCW[2]	PCU[3]
E_M	**0.770**	0.708	0.725	0.755	0.738	0.697
ASY-H[4]	**0.743**	0.702	0.679	0.716	0.700	0.683

[1]*Average of PCN and PCS.*
[2]*Selection of winter hemisphere PC indices.*
[3]*Selection of summer hemisphere PC indices.*
[4]*Magnetic storm events (1992–2018).*

Table 5.
Correlation coefficients for epoch 1998–2018 [from [29]].

(negative) peak values, relates to the intensity-time history of the PC indices and not to any specific instantaneous value.

10.6 Quality control

All the aspects of the investigations of the relations between the polar cap indices and the merging electric field and the ring current properties rely critically on the quality of the basic magnetic data and their proper handling. Thus, with profound respect for such concerns, all magnetic observational data involved here have been inspected in plots like the diagrams displayed in **Figure 1a** and **b** of monthly and yearly averages for international quiet (QQ) days. The component base lines have been controlled and—if needed—corrected to provide smooth secular variations only. The QDC values needed for defining the magnetic variations have also been displayed in yearly summary plots corresponding to the samples presented here in **Figure 2a** and **b** or in **Figure 9** of Stauning [16] defining the solar rotation weighting (SRW) QDC method or the PC index reports [14, 17]. The PC indices have been inspected in monthly plots similar to the samples presented in Stauning et al. [14] or Stauning [17]. In addition, the derived equivalent Dst indices (Dst_{EQ}) have been displayed along with the published Dst indices for each of the storm cases in the formats of **Figures 14** and **16**. Diagrams for the entire integration interval from 1992 to 2018 are included in Appendix A of [29]. It appears that corresponding quality measures have not been considered for the IAGA-endorsed indices [20].

11. Summary

11.1 Relations between the merging electric fields and PC indices

The examinations of the relations between the solar wind merging electric field, E_M, and the polar cap indices, PCN, PCS, and PCC, presented in **Figures 3** and **4** have demonstrated, unambiguously, that the non-negative combined PCC indices present higher values of the correlation with E_M than either of the PCN and PCS indices, their averages (PCA), or the seasonal selections, PCW for the local winter, and PCU for the summer index values throughout all years of available data and throughout all seasons of the years. A summary of correlation coefficients is shown in **Table 5**. On top of the PCC-E_M correlation with Rx = 0.770 shown in **Table 5**, the PCCD-E_M correlation (using Dome-C for questionable Vostok data) provides Rx = 0.786, the best score of all PC index combinations.

The examinations have also shown that the correlations between E_M and PCN or PCS have considerable seasonal variations with minima at local summer conditions, as displayed particularly clear in **Figure 4**. These minima are probably related in part to the corresponding maxima in the intensities of reverse convection (DP3) events. The seasonal variations have been mitigated in the PCC index version leaving a weak minimum at northern winter conditions (cf. **Figure 4**).

11.2 Relations between auroral activity and PC indices

The high, albeit seasonally varying, correlation coefficients and the regression slopes displayed in **Figure 7** provide a useful reference for the relations between polar cap and the auroral current systems. The rather broad range of PC index values

(0.25–2.25 mV/m) preceding substorms with onset steps of –100, –200, or –300 nT displayed in **Figure 9** indicates that the PC indices are not so useful in prediction of relatively weak substorm events. There is a particular problem with negative PCN and PCS index values since substorms are known to occur during strong NBZ conditions too. On the other hand, **Figure 9** also demonstrates that larger PC index levels of 2–4 mV/m are needed to generate the violent substorms with step amplitudes beyond 300 nT, which are those of particular relevance for space weather applications such as the prediction of GIC events threatening power grids [32, 33] rather than the hundreds of small substorms occurring every year.

11.3 Relations between ring current and polar cap indices

Fine-tuning of the control parameters and including modifications of the PCC indices for high-level saturation effects [cf. **Table 3** and Eqs. (13) and (14)] were used to calculate equivalent Dst_{EQ} indices by the integration of the source function [Eq. (10)] throughout the entire interval from 1992 to 2018 without any attachment to the real Dst index series. The correlation between the equivalent Dst index series using the source function based on the PCC indices derived from transpolar convection intensities and the real Dst indices based on near-equatorial magnetic observations reached a value of 0.856 at a delay of 45 min (PCC leading). The mean difference between the two series was below 1 nT, the mean absolute difference was below 10 nT, while the RMS difference was less than 13 nT (cf. **Table 4**).

In a simplified version of the small contributions (≈20 nT) from the magnetopause currents, the Dst (or SYM-H) indices could be derived by integration (summations in small steps) of the rate of change defined by Eq. (10) using the parameters from **Table 3** and Eqs. (13) and (14) to provide the relation shown as follows [from [29]]:

$$d(Dst^*)/dt = gradD \cdot PCCeff – Dst^*/\tau \qquad (15)$$

where

$$Dst* = Dst – 20\ nT$$

$$gradD = –4.5\ (nT/h)/(mV/m)$$

$$PCCeff = PCC \text{ if } PCC < 5\ mV/m \text{ or } PCCeff = PCC + 0.6 \cdot (PCC - 5) \text{ if } PCC > 5\ mV/m$$

$$\tau = 5.5\ h \text{ if } Dst < –52\ nT \text{ or } \tau = 7.0\ h \text{ if } Dst > –52\ nT.$$

Contrary to statements in [25, 26, 35], the present work (cf. **Figure 12**) has no indication of particular thresholds in the PC indices for ring current increases or decays. The ring currents monitored through the Dst_{EQ} (or SYM-H_{EQ}) indices start increasing as the PCC indices rise to a positive level, develops with the integrated time history of PCC, as shown, for instance, in Figures (13) and (14), and decays when the PCC-based source function in Eq. (10) [Eq. (10)] turns positive. There are no specific relations between the amplitude ratios or timing of PC maxima and Dst (or SYM-H) minima. The integration of Eq. (10) starting from quiet conditions (Dst ≈ 0) provides equivalent Dst (or SYM-H) index values up to ≈45 min ahead of actual time. Real-time PCC-based Dst index gradient source values provide the actual equivalent Dst (or SYM-H) forward slopes providing reliable ring current estimates at least 1 h ahead helping to decide whether storm conditions are intensifying or decaying.

12. Conclusions

- The present work has provided a systematic assessment of the correlation between various PC index versions used in published works with the merging electric field, E_M, in the solar wind and with ground-based global magnetic indices such as the auroral electrojet indices, AL, AU, SML, SMU, and the ring current indices, ASY-H, SYM-H, and Dst.

- The relations between the polar cap PCC indices, built from non-negative values of the PCN and PCS indices, and the solar wind merging electric field, E_M, are closer with markedly larger correlation coefficients than those found for the relations between E_M and either of the individual PCN or PCS indices, their averages, or the summer or winter hemisphere PC index selections throughout all years and regardless of the season.

- The correlation between the PCC indices and the auroral electrojet indices, AL and AU (or SuperMag SML and SMU), is so high that meaningful equivalent auroral indices could be derived to supplement or eventually replace the real indices at space weather monitoring.

- Substorm onset conditions relate closely to the PCC index level, particularly at the strong events endangering power grids through their GIC effects.

- For the scaling or forecasting of global disturbance conditions, the development of the asymmetric ring currents related to substorm activity could be monitored from the equivalent PC-based ASY-H index values, which are particularly useful, if the real index values are not available. For such applications, PCC indices rather than either of the hemispherical PC indices or other possible PC index combinations should be used to provide accurate and timely indications.

- The direct correspondence between the instantaneous PC index values and ring current Dst or SYM-H index levels or peak values is poor.

- The PC indices relate to the gradients (rate of change) in the symmetric ring current intensities monitored by the Dst or SYM-H indices. Accurate and detailed equivalent Dst_{EQ} or SYM-H_{EQ} values could be derived to replace or supplement the real ring current indices and provide reliable forecasts of the symmetric ring currents up to 1 h ahead of actual time by the integration of the PCC-based source function from any previous known state.

- The close correspondence between real Dst and equivalent Dst index values at the integration throughout 1992–2018 providing correlation Rx = 0.86, mean deviation below 1 nT, and standard deviation less than 13 nT supports the concept of using PCC indices in a Dst_{EQ} source function. The accurate relations between the PCC-based Dst_{EQ} and the real Dst indices have enabled fine-tuning of timing parameters used in models of the ring current and has supported the modification of the PCC index values to counteract saturation effects at high disturbance levels.

- The high correlation and the accurate timing observed in the relations between the PCC indices based on transpolar convection of plasma and embedded magnetic fields and the ring current indices derived from near-equatorial magnetic variations may provide new insight in and improved modeling of the physical processes linking the polar and equatorial geomagnetic disturbance phenomena and help resolving their common origin in the solar wind properties.

- The PC indices provide a great potential for space weather services by enabling monitoring of the input of solar wind energy to the magnetosphere where the energy is used to power disturbance processes such as polar and auroral activity, upper atmosphere heating, substorms, and geomagnetic storms. The PC indices enabling the input energy monitoring are derived from magnetic variations recorded at two oppositely located polar cap observatories only. The PCC indices improve the accuracy over other PC index versions. Using multiple sources for PCN and PCS indices would greatly improve service reliability.

Acknowledgements

The staff at the observatories in Qaanaaq (Thule), Vostok, and Dome-C, and their supporting institutes is gratefully acknowledged for providing high-quality geomagnetic data for this study. The efficient provision of space data from the IMP, ACE, Wind, and Geotail missions by the OMNIweb service, the supply of geomagnetic data from the INTERMAGNET data service center, and the excellent performance of the ISGI and AARI PC index portals are greatly appreciated. The efforts from the many geomagnetic observatories involved in the collection of data for the SuperMag, AE, Dst, SYM, and ASY indices and their data processing team at the University of Maryland, at GFZ data center in Potsdam, and at WDC-C2 in Kyoto are most gratefully acknowledged. The author gratefully acknowledges the collaboration and many rewarding discussions in the past with Dr. O.A. Troshichev and Dr. A.S. Janzhura at the Arctic and Antarctic Research Institute, St. Petersburg, Russia.

Data availability

Geomagnetic data from Qaanaaq, Vostok, and Dome-C observatories were downloaded from the INTERMAGNET data service web portal at http://intermagnet.org. Ring current indices, Dst, SYM-H, and ASY-H, were downloaded from the web portal for World Data Centre WDC-C2 in Kyoto at http://swdcwww.kugi.kyoto-u.ac.jp/dstdir/index.html. Spacecraft data needed to generate the merging electric field values were downloaded from the OMNIweb service portal http://omniweb.gsfc.nasa.gov. SSC and QD data were downloaded from the ISGI data service portal http://isgi.unistra.fr. The SuperMag indices were downloaded from the web service at http://supermag.jhuapl.edu.

The magnetic observatory in Qaanaaq is managed by the Danish Meteorological Institute, while the magnetometer instruments are operated by DTU Space, Denmark. The Vostok observatory is operated by the Arctic and Antarctic Research Institute in St. Petersburg, Russia. The Dome-C observatory is managed by Ecole et Observatoire des Sc iences de la Terre (France) and Istituto Nazionale di Geofisica e Vulcanologia (Italy).

The "DMI" PC index version is documented in the report SR-16-22 [17] available at the web site: http://www.dmi.dk/fileadmin/user_upload/Rapporter/TR/2016/SR-16-22-PCindex.pdf.

Keypoints

1. Demonstration of close relations between solar wind parameters, polar cap (PC) indices, and major magnetospheric current systems

2. Specification of relations between polar cap (PC), auroral (AL, AU and SML, SMU), and ring current (Dst, SYM-H, ASY-H) indices.

3. Development of methods for the application of real-time polar cap (PC) indices in space weather monitoring.

Author details

Peter Stauning
Danish Meteorological Institute, Copenhagen, Denmark

*Address all correspondence to: pst@dmi.dk

References

[1] Dungey JW. Interplanetary magnetic field and the auroral zones. Physical Review Letters. 1961;**6**:47-48. DOI: 10.1103/PhysRevLett.6.47

[2] Obayashi T. The interaction of solar plasma with geomagnetic field, disturbed conditions. In: King JW, Newman WS. Solar Terrestrial Physics. Academic Press, New York; 1967. p. 107

[3] Nishida A. Geomagnetic Dp 2 fluctuations and associated magnetospheric phenomena. Journal of Geophysical Research. 1968;**73**(5): 1795-1803. DOI: 10.1029/JA073i005 p01795

[4] Nishida A, Maezawa K. Two basic modes of interaction between the solar wind and the magnetosphere. Journal of Geophysical Research. 1971;**76**:2254

[5] Dungey, JW. The structure of the exosphere or adventures in velocity space, in: Geophysics, The Earth's Environment. In: DeWitt C, Hieblot J, LeBeau L, pp. 503–550, Gordon and Breach, New York. 1963.

[6] Lockwood M, McWilliams KA. A survey of 25 years' transpolar voltage data from the SuperDARN radar network and the expandingcontracting polar cap model. Journal of Geophysical Research: Space Physics. 2021;**126**: e2021JA029554. DOI: 10.1029/ 2021JA029554

[7] Fairfield DH. Polar magnetic disturbances and the interplanetary magnetic field. COSPAR Space Research. 1968;**VIII**:107. DOI: 10.1029/ JZ071i001p00155

[8] Kuznetsov BM, Troshichev OA. On the nature of polar cap magnetic activity during undisturbed periods. Planetary and Space Science. 1977;**25**:15-21. DOI: 10.1016/0032-0633(77)90113-1

[9] Troshichev OA, Andrezen VG. The relationship between interplanetary quantities and magnetic activity in the southern polar cap. Journal of Planetary and Space Science. 1985;**33**:415-419. DOI: 10.1016/0032-0633(85)90086-8

[10] Troshichev OA, Andrezen VG, Vennerstrøm S, Friis-Christensen E. Magnetic activity in the polar cap–A new index. Journal of Planetary and Space Science. 1988;**36**(11):1095-1102. DOI: 10.1016/0032-0633(88)90063-3

[11] Vennerstrøm S, Friis-Christensen E, Troshichev OA, Andrezen VG. Comparison between the polar cap index PC and the auroral electrojet indices AE, AL and AU. Journal of Geophysical Research: Space Physics. 1991;**96**(A1): 101-113. DOI: 10.1029/90JA01975

[12] Troshichev OA, Janzhura AS, Stauning P. Unified PCN and PCS indices: Method of calculation, physical sense and dependence on the IMF azimuthal and northward components. Journal of Geophysical Research. 2006; **111**:A05208. DOI: 10.1029/ 2005JA011402 Note correction in Troshichev et al. [13]

[13] Troshichev OA, Janzhura AS, Stauning P. Correction to "Unified PCN and PCS indices: Method of calculation, physical sense, and dependence on the IMF azimuthal and northward components". Journal of Geophysical Research. 2009;**114**:A11202. DOI: 10.1029/2009JA014937

[14] Stauning P, Troshichev OA, Janzhura A. Polar cap (PC) index. Unified PC-N (North) index procedures

and quality. DMI Scientific Report, SR-06-04. 2006. Available from: https://www.dmi.dk/fileadmin/Rapporter/SR/sr06-04.pdf

[15] Stauning P. A new index for the interplanetary merging electric field and geomagnetic activity: Application of the unified polar cap indices. Space Weather. 2007;**5**:S09001. DOI: 10.1029/2007SW000311

[16] Stauning P. Determination of the quiet daily geomagnetic variations for polar regions. Journal of Atmospheric and Solar-Terrestrial Physics. 2011;**73**:2314-2330. DOI: 10.1016/j.jastp.2011.07.004

[17] Stauning P. The polar cap (PC) index: Derivation procedures and quality control. DMI Scientific Report SR-16-22. 2016. Available from: https://www.dmi.dk/fileadmin/user_upload/Rapporter/TR/2016/SR-16-22-PCindex.pdf

[18] Stauning P. Multi-station basis for polar cap (PC) indices: Ensuring credibility and operational reliability. Journal of Space Weather and Space Climate. 2018;**8**:A07. DOI: 10.1051/swsc/2017036

[19] Stauning P. Reliable polar cap (PC) indices for space weather monitoring and forecast. Journal of Space Weather and Space Climate. 2018;**8**:A49. DOI: 10.1051/swsc/2018031

[20] Stauning P. The polar cap (PC) index: Invalid index series and a different approach. Space Weather. 2020;**18**(10):1-18. DOI: 10.1029/2020SW002442 (accepted for publication)

[21] Kan JR, Lee LC. Energy coupling function and solar wind-magnetosphere dynamo. Geophysical Research Letters. 1979;**6**(7):577-580. DOI: 10.1029/GL006i007p00577

[22] Milan SE, Carter JA, Sangha H, Bower GE, Anderson BJ. Magnetospheric flux throughput in the Dungey cycle: Identification of convection state during 2010. Journal of Geophysical Research: Space Physics. 2021;**126**:e2020JA028437. DOI: 10.1029/2020JA028437

[23] Janzhura AS, Troshichev OA, Stauning P. Unified PC indices: Relation to the isolated magnetic substorms. Journal of Geophysical Research. 2007;**112**:A09207. DOI: 10.1029/2006JA012132

[24] Troshichev OA, Podorozhkina NA, Sormakov DA, Janzhura AS. PC index as a proxy of the solar wind energy that entered into the magnetosphere: 1. Development of magnetic substorms. Journal of Geophysical Research: Space Physics. 2014;**119**:6521. DOI: 10.1002/2014JA019940

[25] Troshichev OA, Sormakov DA, Janzhura AS. Relation of PC index to the geomagnetic storm Dst variation. Journal of Atmospheric and Solar-Terrestrial Physics. 2011;**73**:611-622. DOI: 10.1016/j.jastp.2010.12.015

[26] Troshichev OA, Sormakov DA. PC index as a proxy of the solar wind energy that entered into the magnetosphere: (3) Development of magnetic storms. Journal of Atmospheric and Solar-Terrestrial Physics. 2018;**180**:60-77. DOI: 10.1016/j.jastp.2017.10.012

[27] Stauning P, Troshichev OA, Janzhura AS. The polar cap (PC) index: Relations to solar wind parameters and global activity level. Journal of Atmospheric and Solar-Terrestrial Physics. 2008;**70**:2246-2261. DOI: 10.1016/j.jastp.2008.09.028

[28] Stauning P. The polar cap PC indices: Relations to solar wind and global disturbances. In: Lazar M, editor.

Exploring the Solar Wind. London ISBN: 978-953-51-0339-4: IntechOpen; 2012. DOI: 10.5772/37359

[29] Stauning P. The polar cap (PC) index combination, PCC: Relations to solar wind properties and global magnetic disturbances. Journal of Space Weather and Space Climate. 2021;**11**(19):1-23. DOI: 10.1051/swsc/2020074

[30] Kappenman J. Geomagnetic storms and their impact on the U.S. power grid. Metatech Report, Meta-R-319. 2010. 197 pp. Available from: https://www.ferc.gov/sites/default/files/2020-05/ferc_meta-r-319.pdf

[31] Pulkkinen A, Bernabeu E, Thomson A, et al. Geomagnetically induced currents: Science, engineering, and applications readiness. Space Weather. 2017;**15**:828-856. DOI: 10.1002/2016SW001501

[32] Stauning P. Power grid disturbances and polar cap index during geomagnetic storms. Journal of Space Weather and Space Climate. 2013;**3**:A22. DOI: 10.1051/swsc/2013044

[33] Stauning P. Using PC indices to predict violent GIC events threatening power grids. Journal of Space Weather and Space Climate. 2020;**10**(3):1-13. DOI: 10.1051/swsc/2020004

[34] Roeder JL, Jordanova VK. Space weather effects and prediction, Ch. 8. In: Jordanova JK, Ilie R, Chen MW, editors. Ring Current Investigations. Elsevier; 2020. pp. 153-244, ISBN: 978-0-12-815571-4. DOI: 10.1016/B978-0-12-815571-4.00008-1

[35] Space environment (natural and artificial)—Operational estimation of the solar wind energy input into the Earth's magnetosphere by means of the ground-based polar cap (PC) index. ISO Report. International Organization for Standardization. ISO Technical Report ISO/TR/23989-2020. 2020. Available from: https://www.iso.org/standard/77565.html

[36] Akasofu S-I. The development of the auroral substorm. Planetary and Space Science. 1964;**12**(4):273-282. DOI: 10.1016/0032-0633(64)90151-5

[37] Akasofu S-I. Physics of Magnetospheric Substorms. Dordrecht-Holland: D. Reidel Publ. Co.; 1977. pp. 190-220

[38] Akasofu S-I. Interplanetary energy flux associated with magnetospheric substorms. Planetary and Space Science. 1979;**27**:425

[39] Chambodut A, Di Mauro D, Schott JJ, Bordais P, Agnoletto L, Di Felice P. Three years continous record of the Earth's magnetic field at Concordia station (Dome-C, Antarctica). Annals of Geophysics. 2009;**52**:15-24. DOI: 10.4401/ag-4569

[40] Di Mauro D, Cafarella L, Lepidi S, Pietrolungo M, Alfonsi L, Chambodut A. Geomagnetic polar observatories: The role of Concordia station at Dome C, Antartica. Annals of Geophysics. 2014;**57**(6):G0656. DOI: 10.4401/ag-6605

[41] Vennerstrøm S. The geomagnetic activity index PC [PhD thesis]. DMI Scientific Report 91-3. Danish Meteorological Institute; 1991. 105 pp. Available from: https://www.dmi.dk/fileadmin/user_upload/Rapporter/SR/1991/sr91-3.pdf

[42] Matzka J, Troshichev OA. PC_index_description_main_document_incl_Appendix_A.pdf. 2014. Available from: DTU Space web portal. doi:10.11581/DTU:00000057

[43] Janzhura AS, Troshichev OA. Determination of the running quiet daily geomagnetic variation. Journal of Atmospheric and Solar-Terrestrial Physics. 2008;**70**:962-972. DOI: 10.1016/j.jastp.2007.11.004

[44] Akasofu S-I. Relationship between the AE and Dst indices during geomagnetic storms. Journal of Geophysical Research. 1981;**86**:4820-4822

[45] Iijima T, Potemra TA. The amplitude distribution of field-aligned currents at northern high latitudes observed by Triad. Journal of Geophysical Research. 1976;**81**:2165-2174

[46] Iijima T, Potemra TA. Field-aligned currents in the day-side cusp observed by Triad. Journal of Geophysical Research. 1976;**81**:5971-5979

[47] Bythrow PF, Potemra TA. The relationship of total Birkeland currents to the merging electric field. Geophysical Research Letters. 1983;**10**:573-576

[48] Gjerloev JW, Hoffman RA. The large-scale current system during auroral substorms. Journal of Geophysical Research: Space Physics. 2014;**119**:4591-4606. DOI: 10.1002/2013JA019176

[49] McPherron RL, Russell C, Aubry M. 9. Phenomenological model for substorms. Journal of Geophysical Research. 1973;**78**(16):3131-3149. DOI: 10.1029/JA078i016p03131

[50] Davis TN, Sugiura M. Auroral electrojet activity index AE and its Universal Time variations. Journal of Geophysical Research. 1966;**71**:785-801. DOI: 10.1029/JZ071i003p00785

[51] Iyemori T, Araki T, Kamei T, Takeda M. Mid-latitude geomagnetic indices "ASY" and "SYM" for 1999, Geomagnetic indices home page. In: Iyemori T, editor. WDC-C2 for Geomagnetism. Kyoto, Japan: Kyoto University. 2000. Available from: http://swdcwww.kugi.kyoto-u.ac.jp/dstdir/index.html

[52] Gjerloev JW. The SuperMAG data processing technique. Journal of Geophysical Research. 2012;**117**:A09213. DOI: 10.1029/2012JA017683

[53] McPherron RL, Chu X. The midlatitude positive bay index and the statistics of substorm occurrence. Journal of Geophysical Research: Space Physics. 2018;**123**(4):2831-2850. DOI: 10.1002/2017JA024766

[54] Sugiura M, Kamei T. Description of the hourly Dst index. In: Iyemori T, editor. Geomagnetic Indices Home Page. WDC-C2 for Geomagnetism. Kyoto, Japan: Kyoto University; 1981. Available from: http://swdcwww.kugi.kyoto-u.ac.jp/dstdir/index.html

[55] Bartels J. The geomagnetic measures for the time variations of solar corpuscular radiation, described for use in correlation studies in other geophysical fields. Ann Intern Geophys Year. 1957;**4**:227-236

[56] Dessler AJ, Parker EN. Hydromagnetic theory of geomagnetic storms. Journal of Geophysical Research. 1959;**64**:2239-2259. DOI: 10.1029/JZ064i012p02239

[57] Sckopke N. A general relation between the energy of trapped particles and the disturbance field near the Earth. Journal of Geophysical Research. 1966; **71**:3125-3130. DOI: 10.1029/JZ071i013p03125

[58] Burton RK, McPherron RL, Russell CT. An empirical relationship between interplanetary conditions and Dst. Journal of Geophysical Research. 1975;

80:4204-4214. DOI: 10.1029/JA080i031p04204

[59] Feldstein YI, Pisarsky VY, Rudneva NM, Grafe A. Ring current simulation in connection with interplanetary space conditions. Planetary and Space Science. 1984;**32**:975-984. DOI: 10.1016/0032-0633(84)90054-0

[60] Jorgensen AM, Spence HE, Hughes WJ, Singer HJ. A statistical study of the ring current. Journal of Geophysical Research. 2004;**109**:A12204. DOI: 10.1029/2003JA010090

[61] Gao Y. On anomalous departures from a linear relation between the polar cap index and its controlling factors in the solar wind and magnetotail. Journal of Geophysical Research. 2012;**117**: A06201:1-A06201:10. DOI: 10.1029/2012JA017721

[62] Stauning P. Transpolar convection and magnetospheric ring currents: Real-time applications of polar cap (PC) indices. Space Weather. 2021;**19**(7):1-25. DOI: 10.1029/2020SW002072

[63] Lyatsky W, Khazanov GV. A new polar magnetic index of geomagnetic activity. Space Weather. 2008;**6**: S06002. DOI: 10.1029/2008SW000382

[64] Tenfjord P, Østgaard N. Energy transfer and flow in the solar wind-magnetosphere–ionosphere system. A new coupling function. Journal of Geophysical Research: Space Physics. 2013;**118**:5659-5672. DOI: 10.1002/jgra50545

[65] McPherron RL, Hsu T-S, Chu X. An optimum solar wind coupling function for the AL index. Journal of Geophysical Research: Space Physics. 2015; **120**(4):2494-2515. DOI: 10.1002/2014ja020619

[66] IAGA Resolution No. 3. 2013. Available from: http://www.iaga-aiga.org/resolutions

[67] Troshichev OA, Janzhura AS. Space Weather Monitoring by Ground-Based Means. Heidelberg: Springer Praxis Books; 2012. DOI: 10.1007/978-3-642-16803-1

[68] Troshichev OA, Sormakov DA. PC index as a proxy of the solar wind energy that entered into the magnetosphere: (2) Relation to the interplanetary electric field E_{KL}. Earth, Planets and Space. 2015; **67**:170. DOI: 10.1186/s40623-015-0338-4

[69] Troshichev OA, Sormakov DA. PC index as a proxy of the solar wind energy that entered into the magnetosphere: (5) Verification of the solar wind parameters presented at OMNI website. Journal of Atmospheric and Solar-Terrestrial Physics. 2019;**196**:105-147. DOI: 10.1016/j.jastp.2019.105147

[70] Stauning P. The polar cap index: A critical review of methods and a new approach. Journal of Geophysical Research: Space Physics. 2013;**118**:5021-5038. DOI: 10.1002/jgra.50462

[71] Janzhura AS, Troshichev OA. Identification of the IMF sector structure in near-real time by ground magnetic data. Annales Geophysicae. 2011;**29**:1491-1500. DOI: 10.5194/angeo-29-1491-2011

[72] Nielsen JB, Willer AN. Restructuring and harmonizing the code used to calculate the Definitive Polar Cap Index. Report from DTU Space. 2019. Available from: https://tinyurl.com/sx3g5t5

[73] Troshichev OA. Polar cap (PC) index. 2011. Available from: http://pcindex.org (see Supported materials)

[74] Troshichev OA. Polar cap magnetic activity (PC index) and space weather

monitoring. Édition Universitaires Européennes. 2017. ISBN: 978-3-8381-8012-0

[75] Stauning P. Comments on quiet daily variation derivation in "Identification of the IMF sector structure in near-real time by ground magnetic data" by Janzhura and Troshichev (2011). Annales de Geophysique. 2013;**31**:1221-1225. DOI: 10.5194/angeo-31-1221-2013

[76] Stauning P. A critical note on the IAGA-endorsed polar cap index procedure: Effects of solar wind sector structure and reverse polar convection. Annales de Geophysique. 2015;**33**:1443-1455. DOI: 10.5194/angeo-33-1443-2015

[77] Troshichev AO, Lukianova RY. Relation of PC index to the solar wind parameters and substorm activity in time of magnetic storms. Journal of Atmospheric and Solar - Terrestrial Physics. 2002;**64**:585. DOI: 10.1016/S1364-6826(02)00016-0

Chapter 4

The Polar Cap Magnetic Activity (*PC* Index) as a Tool of Monitoring and Nowcasting the Magnetospheric Disturbances

Oleg A. Troshichev

Abstract

PC index was originally introduced as a characteristic of the polar cap magnetic activity generated by geoeffective solar wind coupling with the magnetosphere. Subsequent researches showed that the *PC* index follows changes of the solar wind electric field E_{KL} through the field-aligned current system (R1 FAC) responding to variations of the solar wind parameters. Appearance of magnetospheric disturbances is specified by the *PC* index value (with a typical threshold level ~ 1.5 ± 0.5 mV/m) and by the *PC* index growth rate. The disturbance progression strongly follows the *PC* index variations, the intensity of substorms (AL) and magnetic storms (Dst) being linearly related to the *PC* magnitude. In view of these statistically justified relationships, the *PC* index is regarded at present as a proxy of the solar wind energy input into the magnetosphere. A great advantage of the *PC* index application over other methods, based on the satellite measurements, is a permanent on-line availability of information on the magnetic activity in both northern (*PCN*) and southern (*PCS*) polar caps, providing a means for monitoring the magnetosphere state and for nowcasting the magnetic disturbances development.

Keywords: are Solar wind—magnetosphere coupling, magnetospheric field-aligned currents, magnetic activity in polar caps, PC index, magnetopheric substorms, magnetic storms, monitoring, and nowcasting

1. Introduction

The term "solar wind" is referred to flows of low-energy solar plasma including the magnetic field, which is ejected continuously by the Sun's surface. The Earth's magnetosphere is a result of the solar wind impact on the dipole-like geomagnetic field, the form and size of the magnetosphere being determined by the solar wind parameters such as the solar wind velocity Vsw and the solar magnetic field |B| named usually as an interplanetary magnetic field (IMF). It is totally accepted that the solar wind energy incomes into the magnetosphere, the accumulated energy being realized in form of magnetospheric substorms and magnetic storms. Geomagnetic storms are associated with formation of powerful currents flowing around the Earth at the distance of ~3–7 R_E and displayed as a planetary depression of the geomagnetic field (*Dst* variation) revealing the most power at low- and mid-latitudes [1, 2]. Magnetospheric (or magnetic) substorms [3, 4] are characterized by even stronger magnetic disturbances up to 2000 nT, but they are typical of limited auroral zone, where they are displayed as aurora and are accompanied by a variety of phenomena and processes in the auroral ionosphere.

The polar cap magnetic activity is one of the specific manifestations of the solar wind influence on the magnetosphere, which is displayed in the high-latitude region disposed of poleward of the auroral zone. As it was shown in [5], the polar cap magnetic disturbances correlate the best with the solar wind electric field determined by formula of [6] *EKL = Vsw*(By2 + Bz2)1/2sin2θ/2,* where By and Bz are the azimuthal and vertical IMF components and θ is the clock angle between geomagnetic dipole and IMF tangentional component *BT = (By2 + Bz2)1/***2**. The corresponding *PC* index was put forward [7] as a measure of the solar wind electric field E_{KL} coupling with the magnetosphere. The numerous subsequent studies (see [8, 9]) showed that the *PC* index growth determines the development of magnetospheric disturbances, the intensity of the magnetic storm and substorms being related to the *PC* index value. Thus, the *PC* index proved to be strongly responding to the solar wind E_{KL} field changes, and regulating, afterwards, the development and intensity of the magnetospheric disturbance. Based on these experimental results, the International Association of Geomagnetism and Aeronomy (IAGA) approved the *PC* index as "a proxy for energy that enters into the magnetosphere during solar wind-magnetosphere coupling" [10]. The 1-min *PC* index is calculated on-line by magnetic data from near-pole stations Qaanaaq (Thule) in the Northern hemisphere (*PCN*) and Vostok in the Southern hemisphere (*PCS*) beginning in 1997. The unified method for derivation of the *PCN* and *PCS* indices was put forward in [11].

In this chapter the following topics, revealing the *PC* index significance, are examined: mechanisms of the solar wind influence on the polar cap magnetic activity, relationship between the *PC* index and the solar wind electric field E_{KL}, relationship between the *PC* index and magnetospheric disturbances, the *PC* index use for checking the actual state of magnetosphere.

2. Mechanism of the solar wind influence on magnetic activity in the polar caps

Magnetic alterations typical of the polar caps in periods free of magnetic disturbances in the auroral zone (substorms) were found by *Nagata and Kokubun* [12]. They were named DP2 magnetic disturbances [13] to distinguish them from magnetic substorms (DP1). Current systems of DP2 disturbances consist of two vortices with currents flowing sunward in the near-pole region without any peculiarities in the auroral zone, the current intensity is correlated with the southward IMF [14, 15]. As it was shown later [16, 17], the two-vortex DP2 current system is terminated by the geomagnetic latitudes Φ = 50–60° with the focuses located at the morning and evening poleward boundaries of the auroral oval, and the DP2 disturbances can be observed in the polar cap in absence of the southward IMF.

Other types of magnetic disturbances typical of the sunlight summer polar cap are the "near-pole DP variation", named as DP3 disturbances, which are observed under conditions of northward IMF [16, 18], and magnetic disturbances related to azimuthal IMF [19–21], named as DP4 disturbance [16]. The DP3 system consists of two vortices with opposite anti-sunward directed currents in the very limited near-pole area. The DP4 system includes currents flowing along geomagnetic latitudes with maximal intensity in the daytime cusp region (Φ ~ 80°), the current direction being determined by sign of the IMF azimuthal component.

Figure 1 shows current systems of DP2, DP3, and DP4 magnetic disturbances, generated under action of the southward B_{ZS} (a) and (b), northward B_{ZN} (c), and azimuthal B_Y (d) IMF components [16]. The multi-functional analysis of relationships between the IMF and geomagnetic variations has been fulfilled by *Troshichev and Tsyganenko* [17] to separate effects of the IMF Bx, By, Bz components in case of their combined influence. Results of the analysis have also demonstrated availability of the DP2, DP3, and DP4 current systems associated with action of the southward, northward, and azimuthal IMF components, respectively. The electric field structure

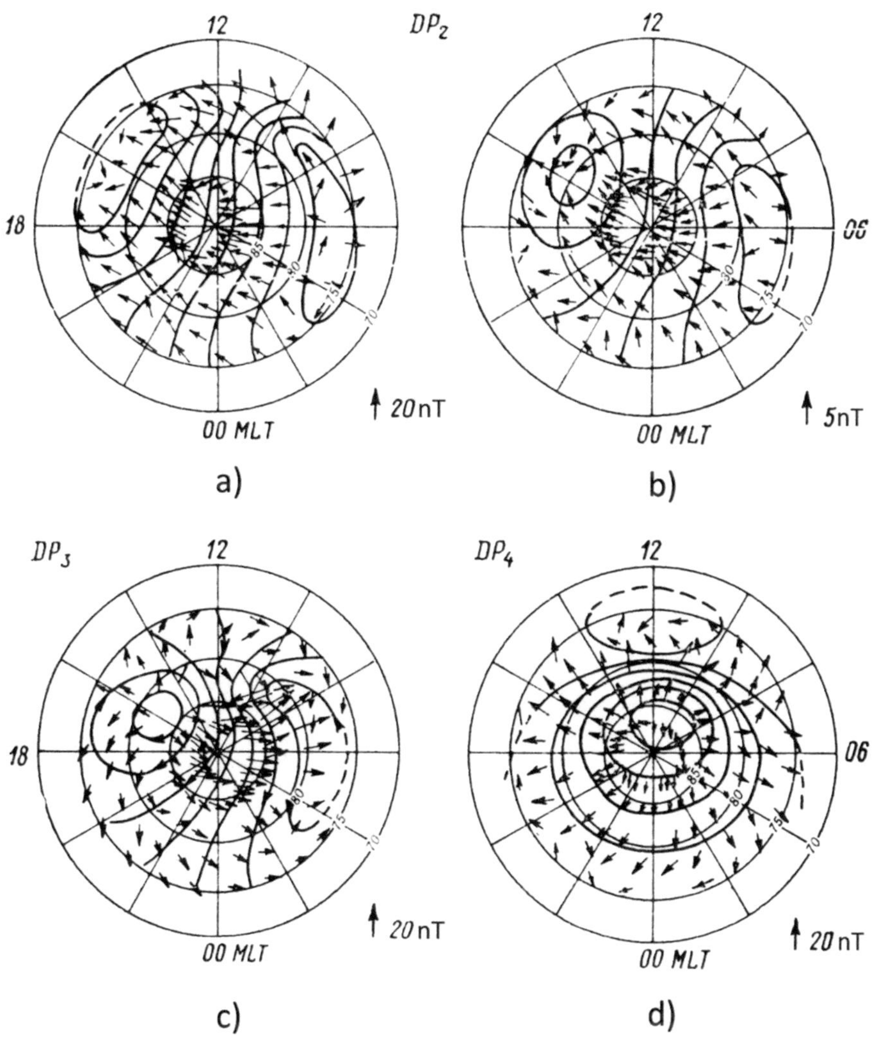

Figure 1.
Current systems of DP2, DP3, and DP4 disturbances related to variations of IMF components: (a) southward B_{ZS} = -1nT, (b) southward B_{ZS} = -0.25nT, (c) northward B_{ZN}, (d) azimuthal B_Y [17].

and intensity derived from magnetic DP2 and DP3 disturbances [16] turned out to be in total agreement with results of direct measurements of electric fields at satellite OGO-6 [22] and in balloon experiments [23].

Mechanism of generation of the polar cap magnetic disturbances became clear when the field-aligned magnetospheric currents were detected onboard the OGO 4 spacecraft [24] and Triad spacecraft [25, 26]. These experiments have fixed a layer of the field-aligned currents on the poleward boundary of the auroral oval (Region 1 FAC system), with currents flowing into the magnetosphere in the morning sector and flowing out of the ionosphere in the evening sector, and layer of the field-aligned currents on the equatorward boundary of the auroral oval (Region 2 FAC system), with opposite directed field-aligned currents. The currents in Region 1 are observed permanently, even during the quiet conditions, whereas Region 2 currents are detected only in periods of magnetic disturbances (*Iijima and* [27]). The intensity of the field-aligned currents demonstrates the strong dependence on the IMF BY and BZ components and the solar wind electric field [28, 29]. During substorm events, the average latitude width of Regions 1 and 2 increases by 20–30% and complicated small-scale structures are superimposed upon the large-scale field-aligned currents, especially in the nighttime sector (**Figure 2**).

The field-aligned currents of reverse polarity were found [30] in the near-pole area, at latitudes of $\Phi > 75°$, under conditions of the IMF northward component (not shown in **Figure 2**). Later these currents were named as NBZ FAC system [31, 32]. The specific BY FAC system, controlled by the azimuthal B_Y IMF component, was separated in the daytime cusp region [33–35]. This FAC system consists of two current sheets located on the equatorward and poleward boundaries of the cusp, the current directions and intensity being determined by the IMF B_Y sign [34, 36]. Influence of the BY FAC system strongly distorts the effects of the regular R1 and NBZ FAC patterns.

It should be noted that R1 and R2 FAC systems presented in [24–26] were outlined by the poleward and equatorward auroral oval boundaries. The same result was

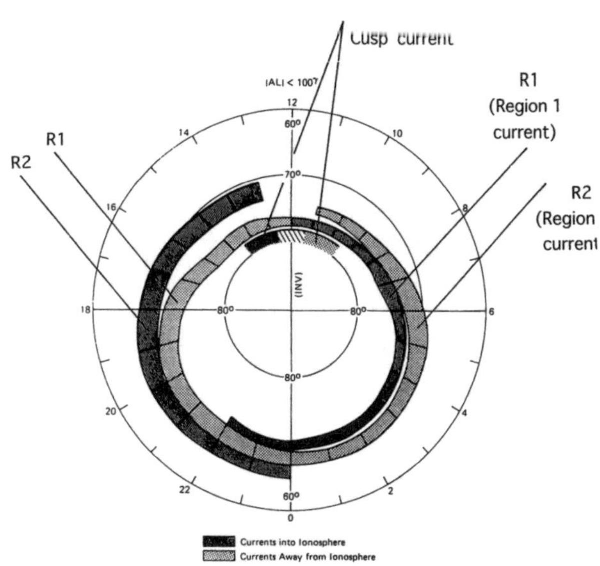

Figure 2.
Pattern of field-aligned currents derived from Triad data [25].

obtained by [37] by measurements onboard the Viking and DMSP-F7 satellites and by [38] by measurements onboard the ISEE 1 and 2 satellites. It implies that generators of R1/R2 FAC systems are positioned within the closed magnetosphere, not on the day-side magnetopause. Results of the R1/R2 FAC mapping to the equatorial plane [27, 39] have also demonstrated that R1 and R2 field-aligned current systems are located within the closed magnetosphere. Availability of the appropriate plasma pressure gradients in the closed equatorial magnetosphere has been displayed in [40, 41].

The numerical simulations of ionospheric electric field and currents generated by field-aligned currents were fulfilled in [42, 43] with use of satellite data [25, 26] on the FAC intensity and structure and data on ionospheric conductivity in the polar caps. The results of numerical simulations have clearly demonstrated that DP2, DP3, and DP4 magnetic disturbances in the polar caps are generated by the corresponding R1, NBZ, and BY FAC systems, the R1 FAC system being presented constantly irrespective of the IMF B_Z polarity. As this takes place, magnetic effect of the ionospheric Pedersen currents in the summer polar cap with high-conductive ionosphere is roughly compensated by the distant magnetic effect of the field-aligned currents, as a result, the magnetic disturbances distribution is determined by ionospheric Hall currents, in full agreement with the theorem of *Fukushima* [44]. On contrary, in the winter polar cap with the low-conductive ionosphere, effect of the ionospheric Hall and Pedersen currents is insignificant, and the polar cap magnetic disturbances are determined by the distant effect of the field-aligned currents. The conclusion was made [45] that the field-aligned currents are responsible for generation of magnetic activity in the polar cap.

Relationship between the *PC* index and really observed field-aligned currents under concrete conditions was examined in [46] based on measurements onboard the SWARM satellites. The analysis, carried out for growth phase of isolated substorms started against the background of magnetic quiescence, showed that increase of the R1 FAC intensity in dawn and dusk sectors of the auroral oval was always accompanied by the *PC* index growth. On contrary, correlation between the *PC* index and field-aligned currents in the noon and the midnight sectors of the oval during the substorm growth phase was absent. In paper [47] relationship between the R1 field-aligned currents and *PC* index was examined for different types of magnetospheric disturbances. The high correlation between the *PC* index and FAC was found, with zero time lag, for all examined events. Thus, the experimental results are indicative of the magnetospheric field-aligned currents as a driver of the polar cap magnetic activity.

3. Response of *PC* index to the E_{KL} field changes

Comprehensive analysis of relationships between the E_{KL} field and *PC* index in course of isolated and expanded substorms (see Section 3) observed in 1998–2001 was fulfilled by *Troshichev and Sormakov* [48] with use of the 1-min E_{KL} and *PC* values. To exclude the possible effect of inconsistency between the “estimated” E_{KL} field (that calculated by solar wind parameters fixed far upstream of the magnetosphere) and actual E_{KL} field affecting the magnetosphere, in reality, the specific “coordinated” substorms (N = 261) were also examined, when the *PC* index and E_{KL} field demonstrated the obviously corresponding variations on the 2-hour interval preceding the substorm onset (SO). The E_{KL} field raise commencement was taken as a key date (T_0) and correlation between the E_{KL} and *PC* quantities over the time

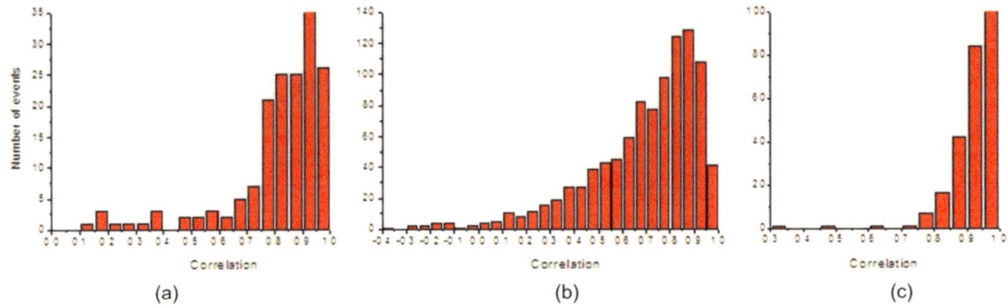

Figure 3.
Histograms of the substorm occurrence over the level of correlation between E_{KL} *and* PC *for isolated (a), expanded (b) and coordinated substorm events (c) [48].*

period $T_0 \pm 30$ minutes was analyzed. **Figure 3** shows distribution of the number of substorms with different coefficients of correlation (R) between the E_{KL} field and *PC* index for various types of substorm events. For coordinated substorms, the correlation was so high as R > 0.7 in 98% of events, the delay time in response of *PC* to E_{KL} field alterations being extended in the range from 0 to 40 minutes with the pronounced peak at ΔT = 10–20 minutes.

To ascertain possible influence of the solar wind parameters on the value of ΔT, the relationships between ΔT and such solar wind parameters as the IMF vertical (B_Z), azimuthal (B_Y) and horizontal (B_T) components, the solar wind speed (*Vsw*) and solar wind dynamic pressure(*Pd*) were examined, the solar wind parameters being averaged for the interval $T_0 \pm 30$ min. Contrary to the expectations, any single solar wind parameter demonstrated a minor importance in the ΔT value setting. In case of the solar wind speed (R = -0.32), only a slight tendency for the decrease of ΔT value with the growth of Vsw was seen. Other solar wind parameters, such as the vertical, azimuthal, and tangential IMF components, did not show any relation to the ΔT value at all.

To reveal the solar wind parameter actually controlling the ΔT value, the 1-min values of V_X, B_Z, E_{KL} and *PC* fixed in course of coordinated events were smoothed with the use of the boxcar average of the 15-min width, and then they were separated into different groups according to delay value ΔT. The smoothed values of V_X, B_Z, E_{KL}, and *PC* were put afterwards through the superposed epoch analysis, the moment of the sudden jump of the 15-min smoothed E_{KL} being taken as a zero time T_0. The behavior of the smoothed values of V_X, B_Z, E_{KL}, and *PC* in course of coordinated events is shown in **Figure 4** for the most statistically justified groups with ΔT = 10–12 min (N = 33), ΔT = 13–15 min (N = 53), ΔT = 16–18 min (N = 60) and ΔT = 19–21 min (N = 38). Thin red lines represent the time evolution of V_X, B_Z, E_{KL}, and *PC* in course of individual events. Solid black lines show the behavior of the mean V_X, B_Z, E_{KL}, and *PC* quantities for each ΔT group. Vertical lines mark the delay time interval boundaries T_0 and $T_0 + \Delta T$, the latter corresponds to the moment when the *PC* index starts to increase.

As **Figure 4** demonstrates, the solar wind speed by itself is not a decisive factor in the ΔT setting (1st panel): the speeds values, as large as V_X = -800 km/s and as small as Vx = −300 km/s, are common for any ΔT group and the time evolution of V_X is not responsive to the moment T_0. The IMF vertical B_Z component (2nd panel) starts to turn down (southward) just at the moment T_0; however, the larger ΔB_Z values are built up at the expense of positive (northward) B_Z IMF preceding the moment T_0, as a result, the delay times ΔT turn out to be shorter under conditions

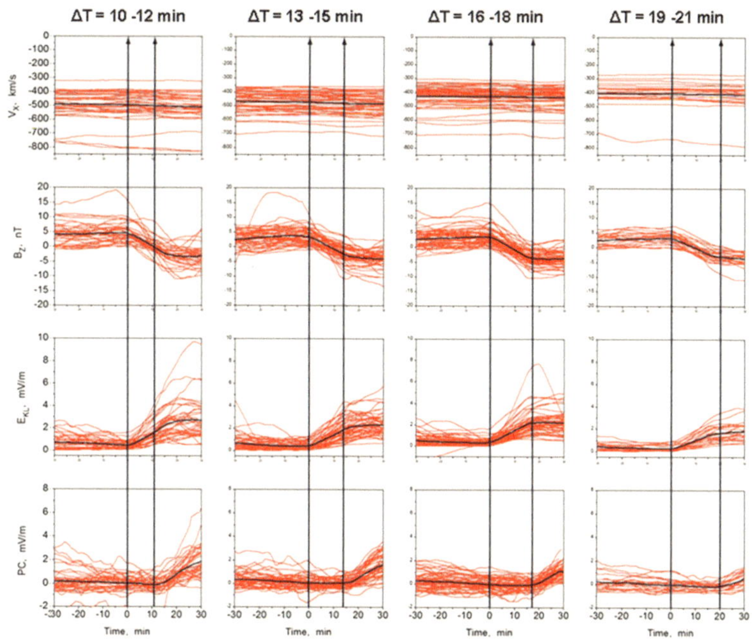

Figure 4.
Time evolution of the 15-min smoothed values V_X, B_Z, E_{KL} *and* PC *observed in case of coordinated events with delay times* ΔT = *10–12, 13–15, 16–18 and 19–21 min. [48].*

of the northward IMF (averaged for the 1-hour interval). At the same time, the correlation between ΔT and the E_{KL} field (3rd panel) turns out to be quite explicit: the higher the E_{KL} raise (ΔE_{KL}) during the ΔT interval is, the shorter the delay time ΔT is. It means that the actual delay time in response of *PC* index to changes of E_{KL} field is determined by the E_{KL} growth rate, not by such solar wind parameters, as the IMF Bz component or the solar wind speed Vsw, contrary to the concept of *Dungey* [49].

4. *PC* index as an indicator of the magnetospheric substorms development

Energy and dynamics of magnetic substorms are commonly estimated by *AL* index, which characterizes intensity of the negative magnetic deviations produced by westward ionospheric currents (auroral electrojets) in the auroral zone [3, 50]. In study [51], the 1-min *PCN* and *PCS* indices, calculated by the unified method [11], were used in analysis of substorms observed in 1998–2001 (N = 1798), the substorm sudden onset (SO) being identified as the *AL* increase by the value more than −100 nT within 15 minutes. It has been demonstrated that the development of magnetic substorms is always preceded by the *PC* index growth. If the *PC* index increases gradually and slightly for a long time, the *AL* index also slowly increases but without SO signatures. The substorm sudden onsets were related to a sharp increase of the *PC* growth occurring within the 10 min interval proceeding the SO moment. Usually, the *PC* index continues to grow after the substorm's sudden onset, the *PC* growth rate being unaffected by SO. The substorm occurrence sharply increases when the *PC* index exceeds the threshold level ~ 1 mV/m and reaches the maximum when *PC* ~ 1.5 mV/m, irrespective of the substorm growth phase duration and type of substorm. Fall of *PC* value below the threshold level leads to substorm completion.

The following classes of magnetic substorms were selected in [8, 51]: **isolated** substorms (N = 194) – disturbances, which arise out of the background of quiet conditions ($AL \leq 200$nT) lasting as a minimum during three hours prior to substorm sudden onset; **expanded** substorms (N = 1418) – disturbances which occurred against the background of noticeable magnetic activity in both the auroral zone and the polar cap; **delayed substorms** (N = 154) – disturbances with sudden onset occurring against the background of invariable, over a long time, magnetic activity substorms. Examples of isolated, expanded and delayed substorms are presented in **Figure 5.**

The results [8, 51] demonstrated that substorms commonly start when the *PC* index exceeds a certain threshold value, i.e. when the energy input into the magnetosphere exceeds a certain crucial level ("energy storage threshold"). It is very essential that this crucial level, dependent on the *PC* growth rate and the magnetospheric activity grade is not the constant value. If the *PC* index (i.e. solar wind energy input) grows gradually and slowly, the magnetic activity also steadily increases, but without substorm onset. It implies that the magnetosphere adapts to new conditions in case of slight energy input. It can occur at the expense of the higher energy dissipation (for example, in the absence of magnetic substorm the Joule heating in the auroral ionosphere is much higher in the periods of enhanced magnetospheric convection than in the periods of ordinary convection). Under these conditions, the balance between the incoming and dissipating energies is retained, but level of energy that is necessary and sufficient for substorm beginning is gradually raised. The substorm is generated by "jump of energy input" when the solar wind energy incoming into the magnetosphere suddenly exceeds the existing level of storage energy. It means that substorm can start with any level of magnetospheric activity and irrespective of how long the solar wind energy was entering into the magnetosphere, in contrast to "directly driven" and "loading-unloading" concepts of the substorm development [3, 52–55]

In case of minor dissipation, when the threshold level is low, the required excess of the energy input over the "storage" energy is insignificant and intensity of the corresponding magnetic substorm will be weak (isolated "magnetic bays" starting against the background of full magnetic quiescence). In case of major dissipation, when the energy crucial level is high, the required excess of the energy input should be significant and the intensity of magnetic disturbance will be, correspondingly, largest (powerful "sawtooth substorms"). In case, when the *PC* index remains unchangeable for tens of minutes after reaching the threshold level and then sharply raises, the "delayed substorm" are observed. Application of the *PC* index as a proxy of the solar wind energy that entered into the magnetosphere gives grounds for verification of the "threshold-dependent driven mode" in different manifestations of magnetospheric substorms.

Figure 6 shows relationships between the *PC mean* and *AL* values in course of isolated, expanded, and delayed substorms obtained for time intervals before the substorm sudden onset SO (T_0, T_0-5min and T_0-20min) and after sudden onset (T_0 + 5 min, T_0 + 10 min, and T_0 + 20 min) [51]. One can see that the slope coefficient after SO turned out to be twice as much as before SO, as an evident consequence of the aurora particle precipitation leading to the rise of conductivity of the auroral zone ionosphere and formation of powerful westward auroral electrojet during the

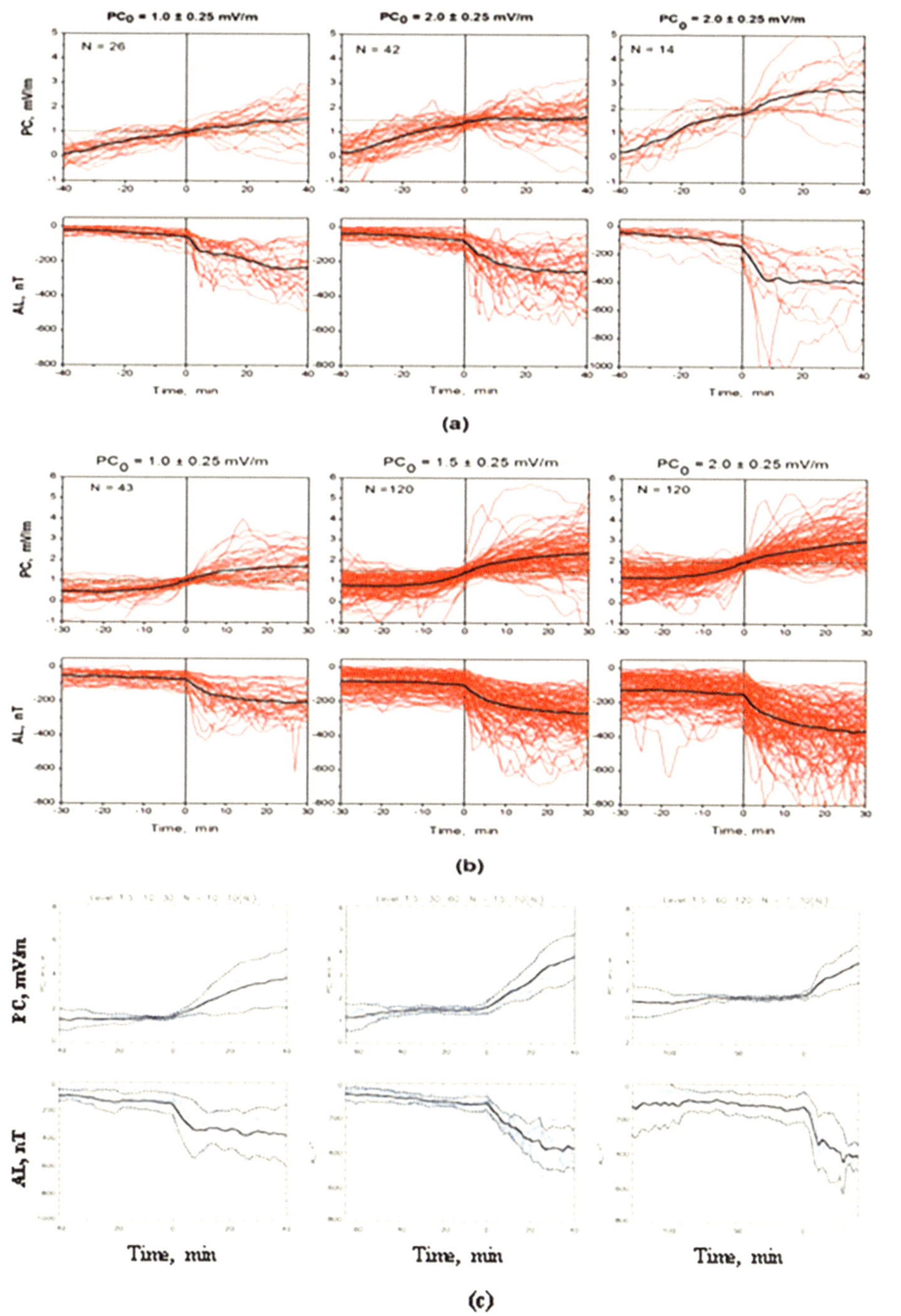

Figure 5.
Examples of isolated, expanded and delayed substorms for different levels of PC *index in moment of substorm sudden onset PC_0 marked by vertical line [51].*

substorm expansion phase. The isolated, delayed, and expanded substorms demonstrate a similar linear dependency of *AL* on *PC* during the substorm expansive phase, with coefficients of correlations changing in range from 0.85 to 0.94.

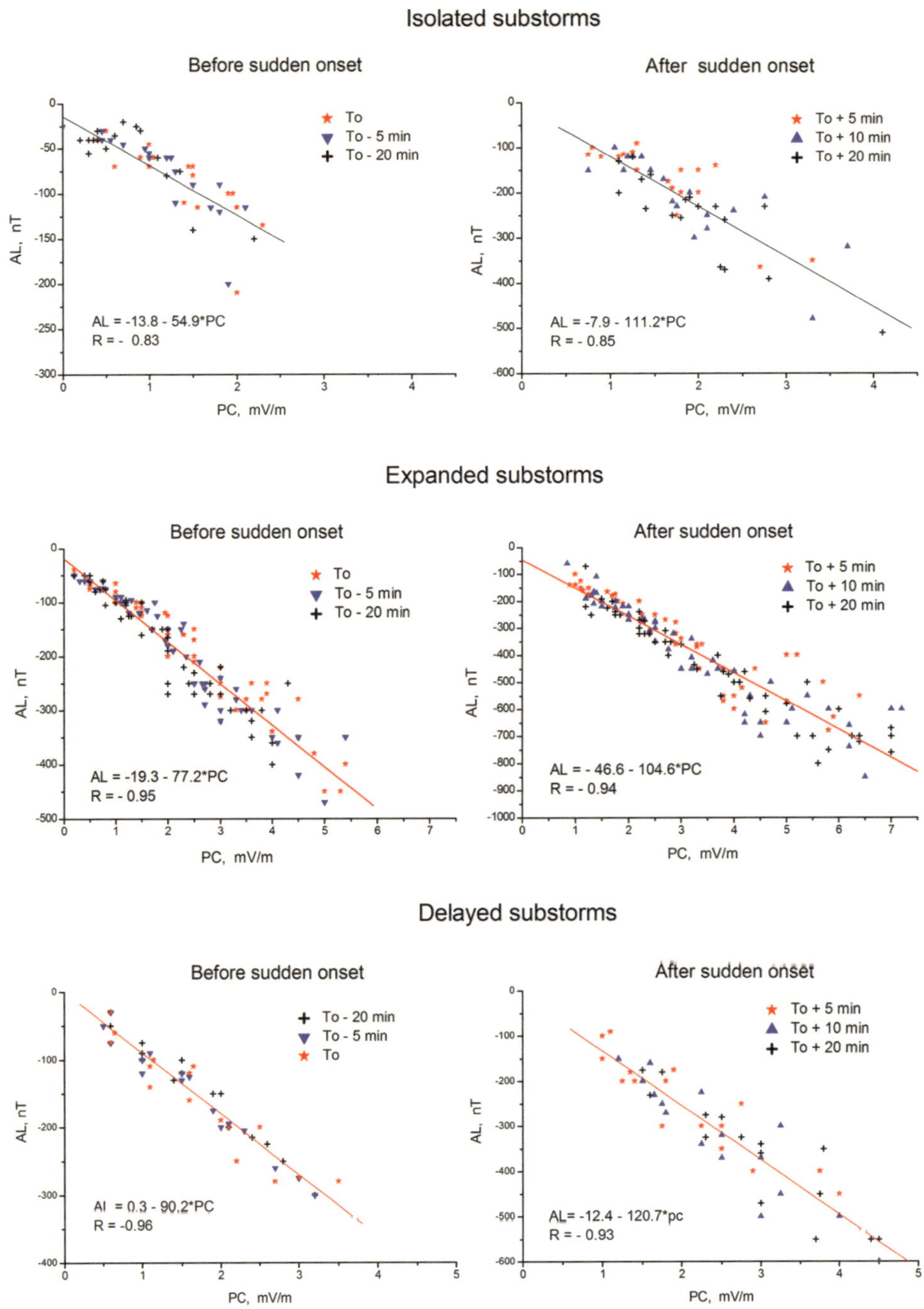

Figure 6.
Relationships between the mean PC *and* AL *values in course of isolated, expanded, and delayed substorms [51].*

5. *PC* index as a precursor of the magnetic storms progression

Term geomagnetic storm is designated for the geomagnetic field depression produced by ring currents flowing in the inner magnetosphere [1]. Intensity of magnetic storms is estimated by 1-hour Dst index [56] or its 1-min analog - *SymH* index. Relationship between the *PC* index and the magnetic storms progression was examined in [57] with the use of the *PC mean* and *SymH* values. The running window of 30-min width was used in the analysis for smoothing the 1-min *PC* and *SymH* indices and E_{KL} field to reveal a link between the processes occurring with quite different time scales (the time scale typical of the E_{KL} field variations and polar cap magnetic activity is ~1 minute, whereas the ring currents are formed in the inner magnetosphere with a time scale of tens of minutes). Results [9] have demonstrated that exceeding the threshold level $PC \sim 1.5$ mV/m is a necessary condition for beginning the magnetic storms like to substorms, but duration of the threshold exceeding should be longer than 1 hour to ensure the storm development. Progression of geomagnetic storms generally follows the time evolution of the 30-min smoothed *PC* index, irrespective of the type and intensity of magnetic storms. Correspondingly, the magnetic storm beginning can be identified as a moment when the PC index steadily exceeds the threshold level. Such identification of the storm beginning turns out to be very fruitful in case of storms with positive DCF effect at the initial storm phase, which is provided by the magnetopause currents responded to the solar wind dynamic pressure.

Three types of magnetic storms were separated in [9, 57] based on peculiarities of the *PC* index behavior, as follows: "**classic storms**", related to ICME impact, with clearly expressed maximum of depression, "**pulsed storms**", related to SIR impact, with periodically repeating oscillations in *PC* and *SymH* indices, and "**combined storms**", which are regarded as the effect of simultaneous ICME and SIR action. **Figure** 7 shows, as an example, the relationship between the *PC* time evolution (upper panel) and the storm progression of *SymH* (lower panel) for different lengths of the storm growth phase in case of classic storms of different intensity (the scale diminution in **Figure 7c** and **d** should be taken into account). Thin red lines show the run of the *PC* and *SymH* indices in course of individual events, thick black solid lines show the behavior of the *PC* and *SymH* values averaged for each storm category and group, the threshold level of $PC = 1.5$ mV/m being marked by the solid horizontal line on upper panels of each figure. The vertical solid lines indicate the moment of the key date $T_0$when the *PC* value steadily (in lapse >1 hour) exceeded the threshold level of $PC = 1.5$ mV/m.

Figure 8 shows relationships between the *PC* and *SymH* values for pulsed storms in category of Dst = −(30–60) nT with roughly constant (a) and decayed (b) amplitudes of *PC* fluctuations, in category of Dst = −(60–90) nT with roughly constant amplitude of *PC* fluctuations (c), and in category of Dst = −(90–120)nT with varying amplitude of *PC* fluctuations (d). The beginning of the pulsed storms is determined, like classic storms, by exceeding the *PC* index above the threshold level $PC = 1.5$ mV/m. However, the further development of storms turns out to be quite different: instead of steady *PC* growth, the pulsed magnetic storms demonstrate the repeated irregular *PC* fluctuations with different periods and different magnitudes (roughly constant, either decayed or alternating) extended over ten hours. The appropriate response in the geomagnetic field depression presents the *SymH* fluctuations of modified periods and smoothed amplitudes, which implies the different processes' actions in the magnetosphere. In order to derive the generalized relationship between the *PC* and *SymH* indices for pulsed storms, the *PC* and *SymH* characteristics, averaged over the main phase duration, were used.

The mean values of *PCmax* and *SymHmin* derived for different categories of storm intensity and growth phase duration (see black solid lines in **Figures** 7 and **8**) were used in [57] to derive a relationship between the appropriate $SymH_{MIN}$ and PC_{MAX} quantities for classic and composite storms. Results of the analysis are presented

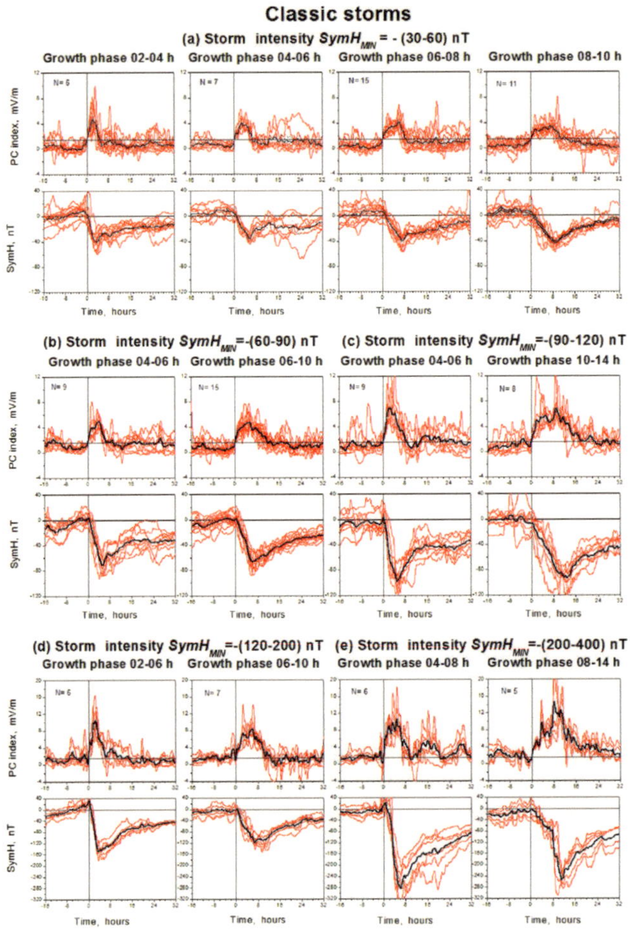

Figure 7.
Relationships between the PC *evolution (upper panel) and the* SymH *progression (lower panel) for 5 categories of classic storms: (a) Dst = −(30-60nT), (b) Dst = −(60-90nT), (c) Dst = −(90-120nT), (d) Dst = −(120-200nT) and (e) Dst = −(200-400nT) with different* PC *growth durations [57].*

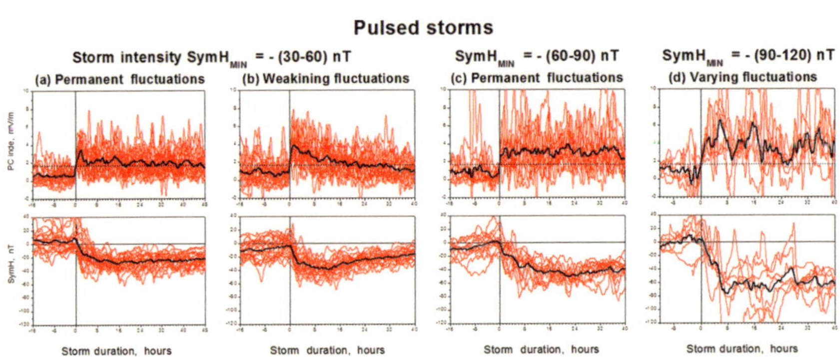

Figure 8.
Relationships between the PC *and* SymH *indices in case of pulsed storms for category* SymH = *−(30–60) nT with roughly permanent (a) and decayed (b) amplitudes of* PC *fluctuations, for category* SymH = *−(60–90) nT with permanent amplitude of* PC *fluctuations (c), and for category* SymH = *−(90–120) nT with alternating PC amplitude (d) [57].*

in **Figure 9** for 26 categories of classic storms (a) and 22 categories of composite storms (b), and for their total (c) with inclusion of the mean data for 4 categories of pulsed storms (olive circles), the standard deviation for each category being marked by vertical bars. The relationships between the storm intensity ($SymH_{MIN}$) and the foregoing PC_{MAX}value are described best by the linear dependences shown in Figure with corresponding correlation coefficients R.

Delay times ΔT in response of $SymH_{MIN}$ to the PC_{MAX} occurrence lie in the range from half-hour to some hours, being dependent on dynamics of the *PC* index alterations. The shortest delay times (ΔT ~ 30–45 min) are observed in case of strong but

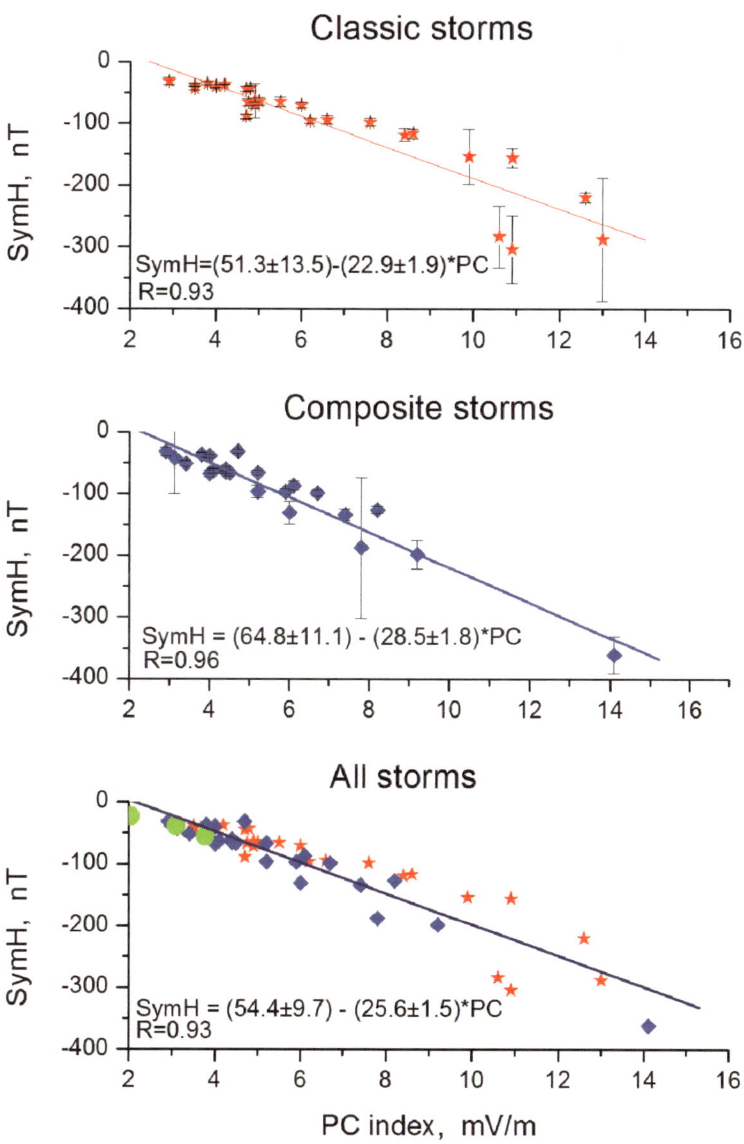

Figure 9.
Relationship between the associated mean values of $SymH_{MIN}$ *and* PC_{MAX} *for classic (a) and composite (b) storms. Panel (c) presents integrated dependence of* $SymH_{MIN}$ *on* PC_{MAX}, *with inclusion of data for pulsed storms (large squares) [57].*

short PC increases when the ring current is quickly formed at a large distance from the Earth and then quickly deceases. The longer delay times are observed in case of the prolonged and irregular PC index dynamics, which initiates the ring currents formation (and subsequent decay) at various distances from the Earth, with the correspondingly different DR current lifetimes. As a result, the longer and unsteady the *PC* growth period is, the larger is delay time.

Thus, the intensity of magnetic storms ($SymH_{MIN}$) is predetermined by value of the PC_{MAX}index, like magnetic substorms. However, it is well to bear in mind that this correspondence was obtained with the use of mean, averaged for 30 minutes, indices for different categories of storms (see **Figure** 7). The actual value of $SymH_{MIN}$ (and delay time ΔT) in each concrete storm event will be depended on the *PC* index dynamics, i.e. on the PC index growth (and decay) rates and duration. It implies that formulas (5)–(7) provide reliable estimations of the storm intensity ($SymH_{MIN}$) for low (< 8 mV/m) *PC* values, the discrepancy between the estimated and actual results being increased while raising the *PC* index value and duration of action.

6. *PC* index as a verifier of the solar wind geoefficiency

In spite of statistically justified agreement in response of E_{KL} field and *PC* index to action of the solar wind, the correspondence between E_{KL} and *PC* can be essentially distorted in the concrete events. The typical examples of consistency and inconsistency between E_{KL} field and *PC* index are presented in **Figure 10**, where the upper panel shows the courses of E_{KL} (green) and *PCmean* = (*PCN* + *PCS*)/2 (violet), the middle panel is for *PCN* and *PCS* indices (blue and red lines), the lower panel shows the *AL*/*AU* indices of magnetic activity (which indicate intensity of negative and positive disturbances in the auroral zone), the substorm onsets being marked by vertical dotted line.

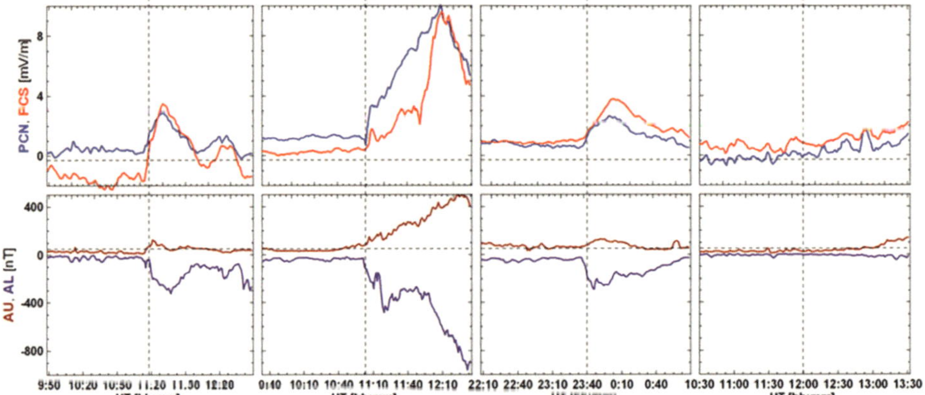

Figure 10.
Examples of consistency and inconsistency in behavior of E_{KL} *field and* PCN, PCS, PCmean *index [58].*

Figure 10a demonstrates concerted changes of E_{KL}, *PC*, and *AL* in course of an isolated magnetic substorm on October 2, 2000, when the disturbance started against the background of quiet magnetic field in response to the *PC* growth related to the E_{KL} field increase. **Figure 10b** demonstrates a specific event on August 17, 2001, when the substorm started in response to the *PC* index jump, but this jump in the ground-based *PC* index was registered ~10 min ahead of the appropriate increase of the estimated E_{KL} field. **Figure 10c** gives example (February 20, 1998), when the electric field E_{KL} was unchanged and quiet (E_{KL} ~ 1 mV/m), whereas the *PCN* and *PCS* indices demonstrated jump above 2 mV/m, which was accompanied, as usual, by the development of substorm with intensity of AL ~ -400nT. In contrast, on 21 October 1999 (**Figure 10d**) the E_{KL} field demonstrated a sharp increase above 2 mV/m for long, but this increase was not followed by the *PC* index growth. It is worthy to note that validity of the *PC* index behavior in all cases was certified by reaction of *AL* index, as a substorm indicator.

It should be reminded that E_{KL} field is estimated by data on the solar wind parameters, such as the solar wind speed Vsw and interplanetary magnetic field (IMF) components, available at the OMNI database (https://omniweb.gsfc.nasa.gov/). These parameters are fixed onboard the spacecraft located far upstream of the magnetosphere, usually at the Lagrange point L1, far upstream of the magnetosphere (at the distance of ~1.5 million km from the Earth). Thereupon they are reduced to the Earth's magnetopause, under the silent presumptions that the solar wind observed in the Lagrange point always encounters the magnetosphere, the Vsw and IMF characteristics is not altered on the way from the L1 point to the magnetopause. That is why the inconsistency between the "estimated" E_{KL} field and *PC* index should be considered as evidence that the solar wind measured by distant monitors did not contact with magnetosphere at all (case of **Figure 10c**), either touched sideways to magnetosphere (**Figure 10d**) or traveled in space with acceleration, as in case of August 17, 2001 (**Figure 10b**), when the real contact of solar wind with magnetosphere (and jump of *PC* index) occurred ahead of the "estimated" contact. As results [58, 59] showed, the solar wind, fixed by distant monitors, did not contact with the Earth's magnetosphere in about 20% of time history and extended in space with acceleration in ~1.5% of examined substorm events. Under these circumstances, the *PC* index takes on great significance as a filter of the OMNI data applicability for analyses of the solar-terrestrial relationship.

7. Relationships between E_{KL} field and *PC*, *AL*, *Dst* indices in 23/24th cycles of solar activity

Invariability of relationships between the E_{KL}, *PC* and *AL*, *Dst* parameters were checked [60] with the use of the daily quantities of these parameters in course of 23/24 solar cycles. **Figure 11** shows, as an example, courses of the E_{KL}, and *PCN*, *PCS* daily values in years of solar maximum (2000, 2015) and solar minimum (2008, 2019). One can see that daily values of E_{KL} and *PC* on average were higher the threshold level 1.5 mV/m in years of solar maximum and lower the level in years of solar minimum, the *PCN*, *PCS* (and *PCmean*) indices being strongly responded to alterations of the E_{KL} field irrespective of the solar activity phase.

To display relationships between the *PC* and *AL*, *Dst* indices, which are quite different in scales of values and changes, the relative values of *PC*, *AL* and *Dst* (i.e. the running quantities related to their average value for period under examination) were taken in [60]. **Figure 12** shows the courses of the relative daily quantities *PCmean*,

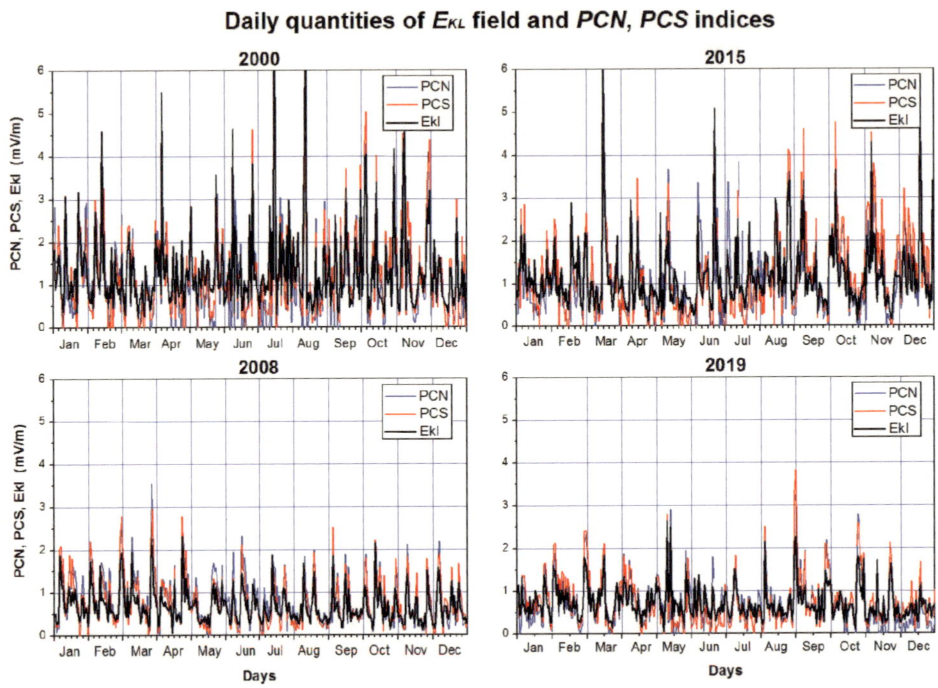

Figure 11.
Coordinated variations of the daily values of E_{KL} *field and* PCN, PCS *indices in epochs of solar maximum (2000, 2015) and solar minimum (2008, 2019) [60].*

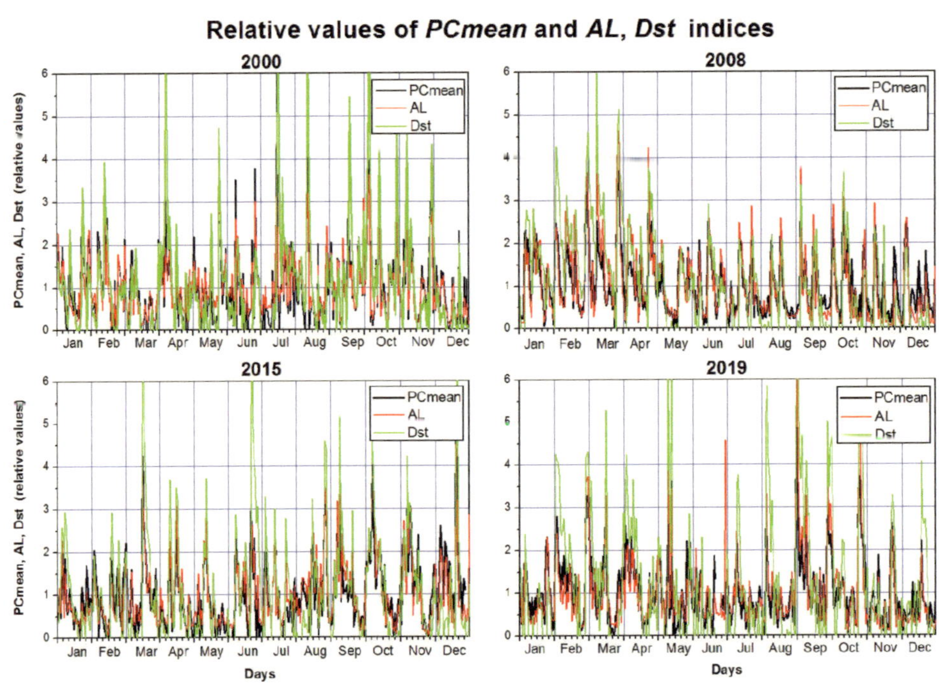

Figure 12.
Coordinated variations of daily values of the PCmean *and* AL, Dst *indices in epochs of solar maximum (2000, 2015) and solar minimum (2008, 2019) [60].*

AL, Dst in 2000, 2015 and 2008, 2019 years, the *PC mean, AL* and *Dst*traces being marked, correspondingly, by black, red, and green colors. It should be reminded in this connection that correlation between the two quantities remains the same irrespective of the kind, absolute or relative, of quantity. One can see that red traces of *AL* index in **Figure 12** practically hide the traces of *PCmean* index. It means that the daily values of *PC* and *AL* indices alter in almost one-to-one correspondence, irrespective of solar activity, which was maximal in 2000, 2015 and minimal in 2008, 2019. The perfect correspondence in courses of the daily *PCmean* and *AL* quantities is consistent with excellent correspondence in variations of the 1-min *PCN, PCS,* and *AL* indices in course of substorm events, noticed in [8, 9].

The daily *Dst* index demonstrates much worse correspondance with *PCmean* than with *AL* index in agreement with reduction of correlation between the corresponding 1-min values while passing from *AL* index to *SymH* index [9]. Indeed, the *AL* index reacts to *PC* index changes in a few minutes, whereas the storm progression responses to the maximal *PC* value with longer and different delay times (from 30 minutes to some hours), as a result, the storm return phase can last for some hours without evident counterparts in the *PC* index (see for example green traces in **Figure 11**, which extended beyond black and red traces). The reason is that the *PC* index dynamics (i.e. by changes in value, duration and rate of the *PC* jumps and drops), determines the different disposition of the DR currents within the magnetosphere and, therefore, the different times of their growth and decay.

As **Table 1** shows, the correlation between E_{KL} field and *PCN, PCS* indices is higher in years of solar minimum (2008, 2019) and worse in years of solar activity maximum (2008, 2015). According to [60] reduction of correlation is related to solar protons (SPE), which intrusion in polar caps extremely increased ionospheric conductivity in the polar caps and violated the regular relationship between the E_{KL} and *PC* values. The daily *AL* index demonstrates the perfect correlation with *PC* irrespective of the solar activity phase.

Year	a	b	R	Year	a	b	R
	PCN = a + b * EKL				PCS = a + b * EKL		
2000	−0.19	1.01	0.85	2000	−0.096	1.04	0.87
2008	−0.08	1.34	0.89	2008	−0.098	1.29	0.91
2015	−0.07	1.02	0.81	2015	−0.16	1.16	0.88
2019	−0.20	1.26	0.87	2019	−0.21	1.33	0.91
1998–2020	0.098	1.106	0.85	1998–2020	0.167	1.161	0.85
	AL = a + b * PCmean				Dst = a + b * PCmean		
2000	−37.9	−86.7	−0.94	2000	−1.6	−14.8	−0.69
2008	9.4	−111.1	−0.93	2008	−1.6	−11.9	−0.72
2015	−18.0	−104.8	−0.93	2015	2.4	−15.1	−0.74
2019	−0.4	−114.5	−0.91	2019	2.6	−11.0	−0.73
1998–2020	−12.35	−100.99	−0.93	1998–2020	1.7	−13.8	−0.72

Table 1.
Relationships between the daily values of E_{KL} *field and* PCN, PCS *indices and between the* PCmean *and* AL, Dst *indices described by linear functions* (Y = a + b*X) *and corresponding coefficients of correlation (R) for years of solar maximum (2000, 2015) and solar minimum (2008, 2019) and for entire period (1997–2020).*

Correlation between the yearly values of E_{KL}, *PC* and *AE*, *Dst* indices in course of 23/24 solar activity cycles was examined [61]. The analysis has demonstrated the remarkable consistency in their variations in 1998–2019, the parameters being perfectly correlated with the yearly values of solar wind velocity *Vsw* and interplanetary magnetic field, as well as with their product electric field E_{KL}. As this takes place, the best correlation of the yearly values E_{KL}, *PC, AE, Dst* was observed with the total IMF field $|B|$ (R = 0.96, 0.84, 0.88, 0.86 correspondingly), not with the IMF Bz component (R = 0.81, 0.72, 0.69, 0.74), in evident contradiction with the Dungey's concept.

Results of analyses [60, 61] indicate that calibration coefficients determining relationship between the E_{KL} field and *PC, AL, Dst* indices remain unchanged during 23 and 24 cycles of solar activity. It means that mechanisms ensuring the solar wind influence on magnetosphere are valid irrespective of solar activity.

8. Discussion

The *PC* index has been affirmed as a proxy of the solar wind energy input into the magnetosphere (Resolution No. 3, IAGA 2013; Resolution No. 2, IAGA 2021). What are the physical implications of this certification? There are three well-known concepts explaining the solar wind's influence on the magnetosphere.

According to the first concept, put forward by [49], the IMF carried by the solar wind contacts with the terrestrial magnetic field at the dayside magnetopause, where the geomagnetic field is northward. When the IMF is southward, the terrestrial field lines will interconnect with the interplanetary field lines, and the electric potential $\Delta V = l_{eff}E$(where E is the interplanetary electric field and l_{eff} is the effective extent of reconnection zone at the dayside magnetopause) will be mapped along infinitely conducting magnetic field lines into the polar ionosphere ensuring the cross-polar cap potential. As a result, the antisunward plasma convection is generated in polar caps, where the ionospheric plasma attached to interplanetary field lines moves together with the solar wind. The merged field lines will reconnect again in the tail neutral sheet, giving rise to the return magnetic flow, which ensures the dayside magneto sphere balance. The Dungey's hypothesis provided the big impulse for the researches of the IMF influence on processes in the magnetospheric. Nevertheless, the hypothesis was criticized from the outset with reference to such theoretical problems as validity of the frozen-in condition in the real magnetosphere, ignoring the turbulence in the real magnetosheath and plasma sheet, necessity to make distinction between the physical laws in passive and active plasma regions and so on.

It should be noted that the original Dungey hypothesis does not even mention the field-aligned currents owing to absence of any information about their existence in those times. At present, the FAC systems registered in the satellite experiments are commonly regarded as favoring the Dungey concept. Indeed, the NBZ FAC system fixed in the near-pole area and BY FAC system fixed in the day-time cusp area [30, 32, 34, 35] can be regarded as a result of interconnection of the interplanetary and terrestrial fields under influence of the IMF northward B_Z and azimuthal B_Y components. However, it is well to bear in mind that these FAC systems are always observed against the background of the permanent R1 FAC system, which continues to exist even under condition of the northward IMF. Moreover, the R1 FACs are positioned far inside the magnetosphere, within the plasma sheet boundaries [39, 62]. The permanent availability of the R1 FAC system (affected by *EKL* field) in presence of independent NBZ (or BY) FAC systems, responding to influence of the northward (or azimuthal) IMF influence, seems to be inconsistent with the interconnection as a reason of the permanent availability of the R1 FAC system. It is possible, that mechanism of interconnection generates, under condition of southward IMF, a specific "SBZ" FAC system (in a similar manner to NBZ and BY systems), but product of this "SBZ" system is added to the effect of permanent R1 FAC system.

The second concept, known as a "viscous-like interaction", was put forward by *Axford and Hines* [63], who suggested that the antisunward plasma convection on closed field lines along the boundary layer of magnetosphere can be ensured via the transfer of the solar wind momentum to the magnetospheric plasma across the magnetopause. At present the viscous-like interaction is regarded as a little effective mechanism, it may be responsible for not more than 15% of the polar cap voltage under the normal solar wind conditions. In addition, it was shown [64] that the particles that originated in the magnetospheric low-latitude boundary layer (LLBL) are positioned in the 09–15 MLT time sector equatorward of day-time cusp, whereas the mantle particles are positioned in the same sector poleward of cusp region. The regularity has been confirmed later by Wing et al. [40]. It implies that the whole boundary layer between the solar wind and magnetosphere plasmas is mapped into the narrow daytime sector, not into the dawn and dusk sectors of the auroral oval, where the R1 FAC system is positioned.

The third concept, formulated ten years later by [65], was elaborated in [66, 67]. According to this concept, the solar wind impact on magnetosphere violates the magnetostatic equilibrium in the outer magnetosphere resulting in the formation of the plasma pressure gradients within the magnetosphere. Redistribution of the plasma pressure leads to generation of large-scale dawn-dusk electric field and initiates the magnetospheric field-aligned currents responsible for cross-polar cap electric potential. The concept of the field-aligned currents generated in the equatorial magnetosphere due to formation of the plasma pressure gradients was supported later by statistically justified data on the plasma gradients distribution in the plasma sheet [40, 41]. Thus, *Tverskoy's* concept predicted, in fact, the existence of the magnetospheric field-aligned currents discovered by [24–26]. Concept of [65] declared that plasma gradients in the magnetosphere are determined by the solar wind impact on the magnetosphere, however, the mechanisms ensuring the link between the solar wind parameters (Vsw and IMF B_Y, B_Z components) and the magnetospheric plasma redistribution have not been defined. As for solar wind dynamic pressure (Psw) influence, it should be reminded that the Psw impulses compressing the magnetosphere lead to magnetic disturbances (polar cap magnetic activity and substorms) only if they are accompanied by the corresponding changes in the E_{KL} field [68].

Thus, the experimental results unambiguously testify that the geoeffective solar wind generates, through the field-aligned currents, magnetic activity in the polar caps. The *PC* index, characterizing the polar cap activity, demonstrates the best relation to the electric field $E_{KL} = Vsw^*(By^2 + Bz^2)^{1/2}sin^2\theta/2$, which is termed as the solar wind electric field. It should be kept in mind that E_{KL} field is displayed only in the Earth's coordinate system (i.e. in the magnetosphere) which is stationary relative to the moved solar wind. Nevertheless, neither of the three above concepts does explain the link between the *PC* index and *EKL* field. What is the *PC* index significance in such a case?

The *PC* index serves as an indicator of capacity of the solar wind influence on the magnetosphere resulting in generation of the electric fields and field-aligned currents responsible for the magnetospheric convection and polar cap magnetic activity and for development of magnetic disturbances. It is very significant that the value and behavior of the *PC* index are related to the solar wind parameters being independent of the intensity and duration of the magnetospheric disturbances. Indeed, features of the *PC* index growth define the onset and intensity of the disturbance, the *PC* index behavior being independent on the disturbance development; to the contrary, the *PC* index decline below the threshold level (~ 1.5 mV/m) is followed by prompt decay of disturbances. This specific feature of the index makes it possible to use the *PC* index not only for monitoring the magnetosphere state but also for nowcasting the disturbance progression.

9. Conclusions

The *PC* index uniquely responds to variations of the interplanetary electric field E_{KL} coupling with the magnetosphere, the E_{KL} effect being transported in the polar caps by means of the field-aligned currents generated in the equatorial magnetosphere due to the plasma pressure gradients. On the other hand, the *PC* index growth predetermines the development of magnetospheric disturbances (substorms and magnetic storms). These experimental results convincingly testify that *PC* index serves as an indicator of the solar wind energy that enters into the magnetosphere during the solar wind-magnetosphere coupling. Advantage of the *PC* index application over other methods, based on the ground-based or satellite data, is permanent on-line availability of information on the magnetic activity in both northern and southern polar caps and, correspondingly, accessibility of information on the solar wind energy input into the magnetosphere. The *PC* index in this charge might be useful for monitoring the space weather, nowcasting the actual state of the magnetosphere, fitting the solar wind-magnetosphere coupling functions, and validating the utility of the solar wind data presented on OMNI website.

A special procedure agreed in 2011 by the Arctic and Antarctic Research Institute (responsible for the production of *PCS* index) and Space Institute of the Danish Technical University (responsible for the production of *PCN* index), ensures the calculation of the 1-min *PC* indices in quasi-real-time based on data of magnetic observations at the polar cap stations Vostok and Qaanaak (Thule). The *PCN/PCS* indices are freely available at the website http://pcindex.org.

Author details

Oleg A. Troshichev
Arctic and Antarctic Research Institute, St. Petersburg, Russia

*Address all correspondence to: olegtro@aari.ru

References

[1] Chapman S, Bartels J. Geomagnetism. Oxford: Clarendon Press; 1940

[2] Cowley SWH. The causes of convection in the Earth's magnetosphere: a review of development during the IMS. Reviews of Geophysics. 1982;**20**:531-565

[3] Akasofu SI. The development of the aurorally substorm. Planetary and Space Science. 1964;**12**:273-282

[4] Akasofu SI. Polar and Magnetospheric Substorms. Dordrecht: Holland; 1968

[5] Troshichev OA, Andrezen VG. The relationship between interplanetary quantities and magnetic activity in the southern polar cap. Planetary and Space Science. 1985;**33**:415

[6] Kan JR, Lee LC. Energy coupling function and solar wind-magnetosphere dynamo. Geophysical Research Letters. 1979;**6**:577

[7] Troshichev OA, Andrezen VG, Vennerstrøm S, Friis-Christensen E. Magnetic activity in the polar cap: A new index. Planetary and Space Science. 1988;**36**:1095

[8] Troshichev O, Janzhura A. Space Weather Monitoring by Ground-based Means: PC Index. Berlin, Heidelberg: Springer Verlag; 2012. p. 288. DOI: 10.1007/978-3-642-16803-1

[9] Troshichev OA. Polar Cap Magnetic Activity (PC Index) and Space Weather Monitoring. Editions universitaires europeennes, ISBN: 978-3-8381-8012-0, p. 140. 2017

[10] XXII IAGA Assembley. 2013. Resolution N 2

[11] Troshichev O, Janzhura A, Stauning P. Unified PCN and PCS indices: Method of calculation, physical sense and dependence on the IMF azimuthal and northward components. Journal of Geophysical Research. 2006;**111**:A05208. DOI: 10.1029/2005JA011402

[12] Nagata T, Kokubun S. An additional geomagnetic daily variation ($S^q{}_p$ field) in the polar region on a geomagnetically quiet day. Reports of Ion Space and Research Japan. 1962;**16**:256-274

[13] Obayashi T. The interaction of solar plasma with geomagnetic field, disturbed conditions. In: King JW, Newman WS, editors. Solar Terrestrial Physics. New York: Academic Press; 1967. p. 107

[14] Nishida A. Coherence of geomagnetic DP2 fluctuations with interplanetary magnetic variations. Journal of Geophysical Research. 1968b;**73**:5549

[15] Nishida A, Maezawa K. Two basic modes of interaction between the solar wind and the magnetosphere. Journal of Geophysical Research. 1971;**76**:2254-2264

[16] Kuznetsov BM, Troshichev OA. On the nature of polar cap magnetic activity during undisturbed conditions. Planetary and Space Science. 1977;**25**:15-21

[17] Troshichev OA, Tsyganenko NA. Correlation relationships between variations of IMF and magnetic disturbances in the polar cap. Geomagn Research. 1979;**25**:47-59

[18] Maezawa K. Magnetospheric convection induced by the positive and negative Z components of the interplanetary magnetic field: Quantitative analysis using polar cap magnetic records. Journal of Geophysical Research. 1976;**81**:2289-2303

[19] Friis-Christnsen E, Lassen K, Wilhjelm J, Wilcox JM, Gonzalez W, Colburn DS. Critical component of the interplanetary magnetic field responsible for large geomagnetic effects in the polar cap. Journal of Geophysical Research. 1972;77:3371-3376

[20] Mansurov SM. A new evidence for relationship between the space and earth magnetic fields. Geomagnetism and Aeronomy. 1969;**9**:768-770

[21] Svalgaard L. 1968. Sector structure of the interplanetary magnetic field and daily variation of the geomagnetic field at high latitudes. Det Danske meteorologiske institute, Charlottenlund, preprint R-6

[22] Heppner JP. Polar cap electric field distribution related to the interplanetary magnetic field direction. Journal of Geophysical Research. 1972;**77**:4877-4887

[23] Mozer FS, Gonzalez WD, Bogott F, Kelley MC, Schutz S. High-latitude electric fields and the three-dimensional interaction between the interplanetary and terrestrial magnetic fields. Journal of Geophysical Research. 1974;**79**:56-63

[24] Zmuda AJ, Armstrong JC. The diurnal flow pattern of field-aligned currents. Journal of Geophysical Research. 1974;**79**:4611-4519

[25] Iijima T, Potemra TA. The amplitude distribution of field-aligned currents at northern high latitudes observed by Triad. Journal of Geophysical Research. 1976a;**81**:2165-2174

[26] Iijima T, Potemra TA. Field-aligned currents in the day-side cusp observed by Triad. Journal of Geophysical Research. 1976b;**81**:5971-5979

[27] Potemra TA. Observations of Birkeland currents with the TRIAD satellite. Astrophysics and Space Science. 1978;**58**:207-226

[28] Bythrow PF, Potemra TA. The relationship of total Birkeland currents to the merging electric field. Geophysical Research Letters. 1983;**10**:573-576

[29] Iijima T, Potemra TA. The relationship between interplanetary quantities and Birkeland current densities. Geophysical Research Letters. 1982;**4**:442-445

[30] McDiarmid IB, Burrows JR, Wilson MD. Reverse polarity field-aligned currents at high latitudes. Journal of Geophysical Research. 1977;**82**:1513-1518

[31] Iijima T, Potemra TA, Zanetti LJ, Bythrow PF. Large-scale Birkeland currents in the day-side polar region during strongly northward IMF: A new Birkeland current system. Journal of Geophysical Research. 1984;**89**:744

[32] Zanetti LJ, Potemra TA, Iijima T, Baumjohann W, Bythrow PF. Ionospheric and Birkeland current distributions for northward interplanetary magnetic field: Inferred polar convection. Journal of Geophysical Research. 1984;**89**:7453

[33] Iijima T, Fujii R, Potemra TA, Saflekos TA. Field-aligned currents in the south polar cusp and their relationship to the interplanetary magnetic field. Journal of Geophysical Research. 1978;**83**:5595-5603

[34] McDiarmid IB, Burrows JR, Wilson MD. Magnetic field perturbations in the dayside cleft and their relationship to the IMF. Journal of Geophysical Research. 1978;**83**:5753-5756

[35] Wilhjelm J, Friis-Christensen E, Potemra TA. The relationship between ionospheric and field-aligned currents in

the day-side cusp. Journal of Geophysical Research. 1978;**83**:5586

[36] Doyle MA, Rich FJ, Burke WJ, Smiddy M. Field-aligned currents and electric fields observed in the regions of the day-side cusp. Journal of Geophysical Research. 1981;**86**:5656-5664

[37] Ohtani S, Potemra TA, Newell PT, Zanetti LJ, Iijima T, et al. Simultaneous prenoon and postnoon observations of three field-aligned current systems from Viking and DMSP-F7. Journal of Geophysical Research. 1995;**100**:119-136

[38] Chan FC, Russel CT. Statistical characteristics of field-aligned currents in the Earth's inner magnetosphere. In: Ohtani S, Fujii R, Hesse M, Lysak R, editors. Magnetospheric Current Systems. Washington: AGU; 2000. pp. 237-243

[39] Antonova EE, Kirpichev IP, Stepanova MV. Field-aligned current mapping and the problem of the generation of magnetospheric convection. Advances in Space Research. 2006;**38**:1637-1641. DOI: 10.1016/j.asr.2005.09.042

[40] Wing S, Newell PT. Quiet time plasma sheet ion pressure contribution to Birkeland currents. Journal of Geophysical Research. 2000;**105**:7793-7802

[41] Xing XL, Lyons R, Angelopoulos V, Larson D, McFadden J, Carlson C, et al. Azimuthal plasma pressure gradient in quiet time plasma sheet. Geophysical Research Letters. 2009;**36**:L14105. DOI: 10.1029/2009GL038881

[42] Gizler VA, Semenov VS, Troshichev OA. The electric fields and currents in the ionosphere generated by field-aligned currents observed by TRIAD. Planetary and Space Science. 1979;**27**:223-231

[43] Troshichev OA, Gizler VA, Ivanova IA, Merkurieva AY. Role of field-aligned currents in generation of high latitude magnetic disturbances. Planetary and Space Science. 1979;**27**:1451-1459

[44] Fukushima N. Equivalence in ground magnetic effect of Chapman-Vestine's and Birkeland-Alfven's electric current systems for polar magnetic storms. Reports in Ion Space Research Japan. 1969;**23**:219-227

[45] Troshichev OA. Polar magnetic disturbances and field-aligned currents. Space Science Reviews. 1982;**32**:275-360

[46] Troshichev O, Sormakov D, BehlkeR. Relationship between PC index and magnetospheric field-aligned currents measured by Swarm satellites. Journal of Atmospheric and Solar: Terrestrial Physics. 2018;**168**:37-47. DOI: doi.org/10.1016/j.jastp.2017.12.020

[47] Adhikari B, Dahal S, Sapkota N, Baruwal P, Bhattarai B, Khanal K, et al. Field-aligned current and polar cap potential and geomagnetic disturbances: A review of cross-correlation analysis. Earth and Space Science. 2018;**5**:440-455. DOI: 10.1029/2018EA000392

[48] Troshichev OA, Sormakov DA. PC index as a proxy of the solar wind energy that entered into the magnetosphere: (2) Relation to the interplanetary electric field E_{KL}. Earth, Planets and Space. 2015;**67**:170. DOI: 10.1186/s40623-015-0338-4

[49] Dungey JW. Interplanetary magnetic field and the auroral zones. Physical Review Letters. 1961;**6**:47

[50] Davis TN, Sugiura M. Auroral electrojet activity AE and its universal time variations. Journal of Geophysical Research. 1966;**71**:785

[51] Troshichev OA, Podorozhkina NA, Sormakov DA, Janzhura AS. PC index

as a proxy of the solar wind energy that entered into the magnetosphere: (1) Development of magnetic substorms. Journal of Geophysical Research, Space Physics. 2014;**119**:6521-6540. DOI: 10.1002/2014JA019940

[52] Baker DN, Akasofu S-I, Baumjohann W, Bieber JW, Fairfield DH, Hones EW, et al. Substorms in the magnetosphere. In: Butler DM, Papadopoulos, editors. Solar-Terrestrial Physics – Present and Future. Washington: NASA; 1984

[53] Kamide Y, Baumjohann W. Magnetosphere-Ionosphere Coupling. Springer-Verlag; 1993

[54] Rostoker G, Akasofu SI, Baumjohann W, Kamide Y, McPherron RL. The roles of direct input of energy from the solar wind and unloading of stored magnetotail energy in driving magnetospheric substorms. Space Science Reviews. 1987;**46**:93

[55] Russell CT, McPherron RL. The magnetotail and substorms. Space Science Reviews. 1973;**15**:205-266

[56] Sugiura M. Hourly values of equtorial Dst for the IGY. Annals of IGY. 1976;**35**:1

[57] Troshichev OA, Sormakov DA. PC index as a proxy of the solar wind energy that entered into the magnetosphere: (3) Development of magnetic storms. Journal of Atmospheric and Solar: Terrestrial Physics. 2018;**180**:60-77. DOI: 10.1016/j.jastp.2017.10.012

[58] Troshichev OA, Sormakov DA. PC index as a proxy of the solar wind energy that entered into the magnetosphere: (5) Verification of the solar wind parameters presented at OMNI website. Journal of Atmospheric and Solar. Terrestrial Physics. 2019b;**196**:105147. DOI: 10.1016/j.jastp.2019.105147

[59] Vokhmyanin MV, Stepanov NA, Sergeev VA. On the evaluation of data quality in the OMNI interplanetary magnetic field database. Space Weather. 2019;**17**:476-486. DOI: 10.1029/2018SW/002113

[60] Troshichev OA, Dolgacheva SA, Sormakov DA. PC index as a proxy of the solar wind energy that entered into the magnetosphere: (6) Invariability of the PC relation to the solar wind electric field EKL and magnetospheric disturbances during 23/24 solar activity cycles. 2022

[61] Troshichev OA, Dolgacheva SA, Stepanov NA, Sormakov DA. The PC index variations during 23/24 solar cycles: Relation to solar wind parameters and magnetospheric disturbances. Journal of Geophysical Research, Space Physics. 2021;**126**:e2020JA028491. DOI: 10.1029/2020JA028491

[62] Antonova EE, Kirpichev IP, Stepanova MV. Plasma pressure distribution in the surrounding the Earth plasma ring and its role in the magnetospheric dynamics. Journal of Atmospheric and Solar: Terrestrial Physics. 2014;**115**:32-40. DOI: 10.1016/j.jastp.2013.12.005

[63] Axford WI, Hines CO. A unified theory of high-latitude geophysical phenomenac. Canadian Journal of Physics. 1961;**39**:1433-1464

[64] Newell PT, Meng CI. Mapping the dayside ionosphere to the magnetosphere according to particle precipitation characteristics. Geophysical Research Letters. 1992;**19**:609-612

[65] Tverskoy BA. Electric fields in the magnetosphere and the origin of trapped

radiation. In: Dyer ER, editor. Solar-Terrestrial Physics. Vol. 3. Dordrecht, Holland: D. Reidel; 1972. pp. 297-317

[66] Antonova EE, Tverskoy BA. On the nature of electric fields in the Earth's inner magnetosphere. Geomagnetism and Aeronomy International. 1998;**1**:9-21

[67] Antonova EE. Magnetostatic equilibrium and turbulent transport in Earth's magnetosphere: A review of experimental observation data and theoretical approach. Geomagnetism and Aeronomy International. 2002;**3**:117-130

[68] Troshichev OA, Sormakov DA. PC index as a proxy of the solar wind energy that entered into the magnetosphere: 4. Relationship between the solar wind dynamic pressure (PSW) impulses and PC, AL indices. Journal of Atmospheric Solar-Terrestrial Physics. 2019a;**182**:200-210. DOI: 10.1016/j.jastp.2018.12.001

Chapter 5

Ionospheric Electron Density and Electron Content Models for Space Weather Monitoring

Wellen Rukundo

Abstract

Monitoring and prediction of space weather phenomena and associated effects requires an understanding of the ionospheric response related to ionospheric electron content and electron density redistribution. These ionospheric response effects to space weather over time have been quantified by ground station measurements (ionosondes, radars, and GPS), satellite and rocket measurements, and estimations from ionospheric models. However, the progressive development of ionospheric models has had inconsistences in trying to describe the redistribution of electron density in response to extreme space weather conditions. In this chapter, we review and discuss the recent developments, progress, improvements, and existing challenges in the developed ionospheric models for prediction and forecasting space weather events and the need for continuous validation. The utilization of deep learning and neural network techniques in developing more flexible, reliable, and accurate data-driven ionospheric models for space weather prediction is also discussed. We also emphasized the roles of International and national Organizations like COSPAR, URSI, ITU, CCIR, and other research and education institutions in supporting and maintaining observatories for real-time monitoring and measurements of ionospheric electron density and TEC.

Keywords: IRI, ionospheric electron density, TEC, ionospheric storms, space weather, equatorial regions, neural networks

1. Introduction

Space weather conditions impact the space environment, the most important for this chapter is the ionosphere of which its variation disrupts the communication links between ground and space systems. The ionosphere plays a vital role in sustaining communication links between space and ground satellite segments applied in aviation, remote sensing, navigation, and surveillance. Also, the variation in the ionosphere affects the propagation of high and medium-frequency radio and TV waves. The latitudinal, daily, and seasonal variation in the ionosphere depends on solar activity and related solar cycle effects and enhanced geomagnetic activity. The ionospheric variation shows repeatedly climatological conditions dependent on solar zenith angle and sun's activity over a long-term period of the solar cycle during quiet solar periods while it depends on space weather dynamics associated with enhanced geomagnetic activity during disturbed solar periods.

The delay effect induced by the ionosphere's dispersive, refractive, and scattering nature on the signal path is proportional to ionospheric Total Electron Content (TEC). Therefore, TEC is a measurement parameter for ionospheric variation calculated from the summation of electron density along any line of sight of a signal. In addition, ionospheric layers (D, E, F1, and F2) varies in altitude ranges and electron density concentrations which respond differently during space weather phenomena. While ionospheric TEC is calculated at F2 peak (350–550 km) from GPS measurements using a thin layer model [1], it's important to note that the plasmasphere at altitudes of 2000 km and above contributes to TEC with approximately 10–60% during the day and nighttime respectively [2]. Therefore, TEC and electron density variations determine the irregular ionospheric conditions with the associated space weather effects.

Other than TEC and electron density models, global TEC maps developed from GPS data provided by International GNSS Services (IGS) connecting a wide network of ground GPS receivers are powerful tools for space weather monitoring (**Figure 1**). Real-time Global Ionosphere Maps (GIMs) and Regional Ionosphere Maps (RIMs) define TEC distribution using several computational algorithms [3, 4] from which several empirical TEC models have been developed. GIMs and RIMs are important tools for validating existing models for performance improvement because a worldwide network of GPS receivers is associated with low-cost implications for maintenance and installation, high accuracy, easy use, continuous operation, accessibility, and high resolution of temporal and spatial variation in GPS-TEC measurements. Other data sources include ionosondes which monitor ionospheric parameters up to F2 peak, Incoherent Scatter Radars (ISR) which measure ionospheric parameters for the bottom and topside ionosphere, rocket measurements of ionospheric parameters for low altitude ionospheric regions, and satellite in-situ measurements of global

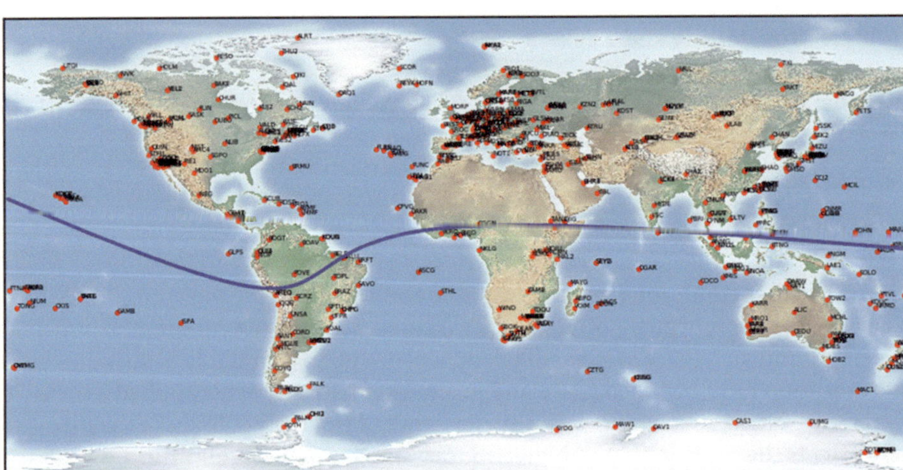

Figure 1.
A map showing worldwide distribution of ground GNSS receiver stations extracted from https://igs.org/network/. The blue line is the geomagnetic equator.

ionospheric parameters important for evaluating model performances and correlation relationship with the measured parameters.

In this chapter, we discuss the latest developments and improvements in ionospheric models for TEC and electron density and the existing challenges. Several models show inconsistencies during storm periods and ionospheric plasma irregularities induced by either latitude or altitude variations. We also discuss the role of deep learning and neural networks techniques in improving space weather prediction capabilities.

2. Ionospheric models for space weather

2.1 Classification of ionospheric models

The ionospheric models are classified as empirical and analytical models (data-driven) and theoretical and parameterized models (physics-based models).

Empirical models are developed from statistical analysis of long-time datasets or combine existing electron distribution models or statistical and numerical analysis of different TEC measurements at local, regional, and global observation and monitoring stations. They define electron density profiles based on geomagnetic/geographical position, solar and geomagnetic activity to produce seasonal and monthly average or median values at specific times and locations. These models mainly reproduce long-term ionospheric response/climatological conditions which depend on the solar activity epoch (diurnal, seasonal, and solar cycle). Their dependence on actual measurements is a challenge in regions with limited and non-uniform distribution of ground and regional TEC measurement and monitoring stations. That's why global empirical models give more representative and accurate TEC estimation in the northern hemisphere and the mid-latitude regions compared to the southern hemisphere [5, 6].

Analytical models are a family of three models (NeQuick, COSTprof, and the ionosphere plasmasphere model) developed from the combination of improved DGR model [7] and diffusive equilibrium models for topside ionosphere and plasmasphere [8]; these apply Epstein functions and spherical harmonics to analytically describe the electron density profiles. The most important analytical model for space weather monitoring is the NeQuick model [9]; this has been adopted by European Geostationary Navigation Overlay System (EGNOS) project for real-time ionospheric corrections of the Galileo single frequency users. In the NeQuick 2 model [9], and the NeQuick model (Galileo Specific) [10], the topside and bottom side parameters have been formulated with modification in electron density, height and thickness parameters and computer package. These models have shown improved performance in TEC estimation during geomagnetic storms when compared with the IRI-2016 model [11]. In addition, the NeQuick 2 model underestimates the topside electron density when compared with IRI-Plas model due to plasmaspheric electron contribution between the 2000 to 20000 km [12], though it performed better than IRI-Plas in TEC estimation from 2013 to 2018 [13]. Generally, the NeQuick model correctly estimates the trend of GPS-TEC during quiet and storm days however the accuracy reduces during geomagnetic storms.

Theoretical models apply first principles to solve an interconnected set of equations including Boltzmann, energy, momentum, and continuity equations. They describe the state of ionospheric plasma along magnetic field lines in phase space with boundary conditions and coupling mechanisms at the lower and upper limits. Magnetospheric, ionospheric, and thermospheric inputs parameters are provided by already developed empirical models for example MSIS-86 [14], HWM-91, or HWM-93 [15], Chiu and IRI empirical models. The accuracy of theoretical models depends on the accuracy, reliability, and representativeness of the empirical models used, the boundary conditions, the fundamental physical and chemical processes, assumptions taken, and the accuracy of numerical schemes. The biggest challenge with the theoretical models is the long time required for processing and computation of the model results which makes it hard to be adopted in the normal operational setting. However, physics-based models provide a more detailed study and explanation of physical and chemical processes about the behavior of ionospheric plasma and coupling mechanisms under different space weather conditions.

Parameterized models [16] were developed by US Air Force Research Laboratories and describe the climatological conditions of the ionosphere by producing electron density profiles, critical frequencies, and heights of E and F ionospheric regions from 90 to 25000 km and TEC. Parameterized models produce climatological conditions from a summarized output of theoretical models for varying geophysical conditions; this is more representative than average ionospheric conditions by empirical models. And hence the driving forces of geophysical conditions associated with winds and wave motion in the lower thermosphere to ionospheric density and TEC variation can be represented.

2.2 TEC and electron density models

Ionospheric models give average modeled TEC and electron density values in space and time, these may deviate from the actual measured values from ionosondes, radars, and GPS measurements by 20–40% during quiet times and can go high during disturbed and geomagnetic conditions. The most efficient models for space weather monitoring must be able to define electron density redistribution caused by both spatial and temporal ionospheric variations. The earliest primary objective behind the development of ionospheric models was to estimate correct ionospheric delay for single-frequency GPS users however, several modifications and improvements with longtime modeled data have enhanced the model's capacity for space weather monitoring and prediction. Their performance in monitoring space weather-induced electron density variations depends on several factors like latitude, altitude, and space weather conditions: hence the development of various regional and altitude-specific models. Global ionospheric models combine several regional and altitude-specific models to give TEC averages globally however their challenge is limited performance on extreme space weather conditions and induced ionospheric plasma irregularities discussed later in Section 3.

Electron density profiles in models are described by ionospheric characteristics, the most important for TEC calculation are peak density (NmF2), critical frequency (foF2), and peak height (hmF2) at the F2 layer where there is maximum electron density. The CCIR and ITU-R maps are standard models for ionospheric parameters set by the International Radio for Consultative Committee (CCIR) and the International Telecommunication Union (ITU). Currently, CCIR maps provide peak parameters for most empirical and theoretical ionospheric models for prediction.

The earlier ionospheric parabolic layer models [17–20] utilized parabolic functions to describe electron density profiles; however, these were challenged by the discontinuities between F1 and E regions, this was later improved by Dudeney [21] by employing trigonometric functions to represent the E-valley (**Figure 2b**).

Chapman layer models utilize the Chapman functions to estimate the rate of photoionization with consideration of one neutral gas species and ignoring transport processes (**Figure 2a**). The neutral density at any height $N(h)$ (Eq. (1)) is determined by the shape of the layer with scale height (H) and peak height (h_m).

$$N(h) = N(h_m)\exp\left(k\left[1 - \frac{h-h_m}{H} - \exp\left(-\frac{h-h_m}{H}\right)\right]\right) \tag{1}$$

The constant $0.5 \le k \le 1$ has been derived empirically in comparison with TEC; this has redefined the height parameters of the Chapman functions representing E, F1, and F2 layers. Distinctively, the α -Chapman layer occurs at $k = 0.5$ while the β -Chapman layer at $k = 1$ [22].

Refer to **Table 1** for details of empirical and semi-empirical electron density and TEC models.

2.2.1 The international reference ionosphere (IRI) model

This is the only standard ionospheric empirical model that takes data from all available data sources. The IRI density profile is based on a combination of global and regional models for D, E, F1 & F2 ionospheric layers with their characteristic parameters merged by mathematical functions. The IRI working group composed of the International Scientific Community, the Committee on Space Research (COSPAR), and the International Union of Radio Science (URSI) updates the model yearly with new data from a network of data sources to validate and improve the model performance. In addition to TEC and electron density, the model describes electron and ion temperatures, and ion composition from an altitude of 50–2000 km at any given time

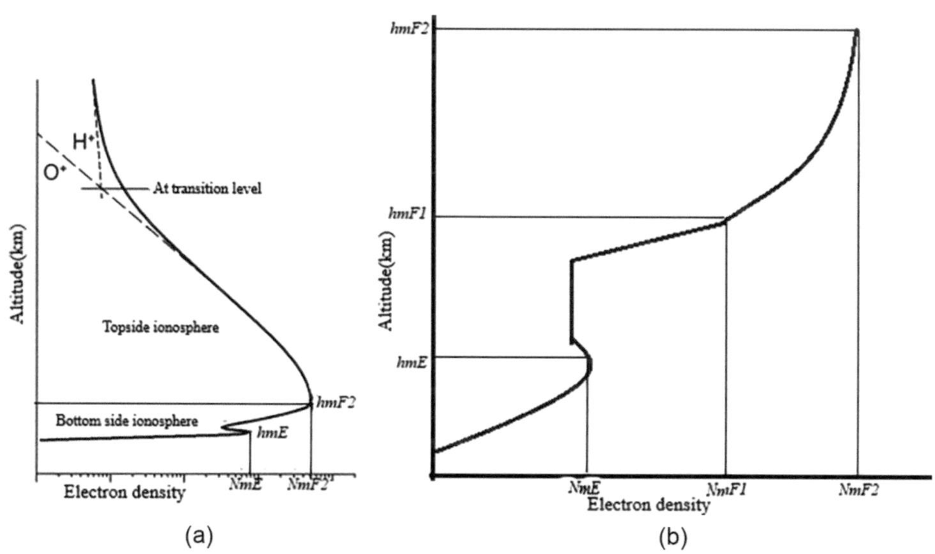

Figure 2.
The representation of electron density models, (a) is a Chapman layer model modified after [22] and (b) is a parabolic layer model modified after [20]. NmE, NmF1, and NmF2 represent peak densities of E, F1, and F2 layers respectively, and hmE, hmF1, and hmF2 represent peak heights of E, F1, and F2 respectively.

Model	Region	Data base	Characteristics
Rush–Miller [23]	Electron density from E to F2 layer	Ionosonde	Chapman, parabolic, CCIR peaks
The Penn State Mk III [24]	Electron density from 120 to 1250 km	Satellite Theoretical computations Empirical models	F-peak parameters
Llewellyn [18]	F region up to 1000 km	Ionosonde, Satellite	Parabolic and exponential segments, CCIR peaks
IRI [25]	Electron density and TEC from 50 to 2000 km	All data sources	Combines multiple models
DGR [7]	Electron density from E to F2 layer	Ionosonde	Epstein functions using analytical expressions
NeQuick [9] NeUoG-plas [26]	Electron density and TEC from E layer to Plasma-sphere	Ionosonde	Epstein functions and plasmaspheric models
Chiu [27]	Electron density from E to F2 layer	Ionosonde	Modified Chapman functions for E-, F1-, and F2-layers
Koehn-lein [28]	Electron density from 50 to 4000 km	Incoherent Scatter Radar, Satellites	Spherical harmonics
Semi-Empirical Low-Latitude Ionospheric Model (SLIM) [29]	Electron density from 180 to 1800 km	Incoherent Scatter Radar, and theoretical simulation from other models	Modified Chapman function
Ionospheric Conductivity and Electron Density (ICED) [30]	Electron density from E to F2 layer	Ionosonde, Satellite	Chapman profiles
Ionospheric Communication Analysis and Prediction (IONCAP) [17]	Below F2	Ionosonde	Parabolic, linear, exponential, CCIR peaks
RIBG [31]	Electron density and TEC from E to F2 layer	Several models	ICED, Bent, plasmasphere model

Table 1.
A table of empirical and semi-empirical ionospheric models for TEC and electron density.

and location. The input parameters to the model represent solar activity, ionospheric index, and magnetic activity and over the years, the model has been updated to the current IRI-2016 [25].

Modeling the topside profile of the electron density has been an existing challenge in the IRI model. This IRI topside has been formulated with a correction factor and the NeQuick model was adopted as a default topside model up to approximately 2000 km [6]. However, this is still not comparable to the altitude approximation of

GPS-TEC at about 20200 km (geostationary orbit altitude). In the latest update of the IRI model (IRI-2016) [25], the representation of the topside ion densities during very low and high solar activity has been improved and the computer program updated to turn IRI into a real-time model for space weather forecasting. The extension of the IRI model to the IRI plasmasphere model (IRI-Plas) [32] has been proposed to improve the representation of the plasmaspheric electron contribution. However, model results have shown overestimation when compared with actual measurements from GPS and ionosondes on varying space weather conditions [33, 34]. This, therefore, suggests that the IRI-Plas extension needs further improvement and update to correctly estimate the plasmaspheric electron content with response to dynamic space weather. Despite the regular updates and improvements, limited data contribution to yearly IRI updates from a limited number of ionospheric monitoring stations in African region (**Figure 1**) in addition to the equatorial anomaly irregularities is the most cause of significant bias between IRI modeled results and experimental GPS-TEC [35]. NeQuick being a 3-dimensional and with IRI models are the only recommended models for calculation of TEC by Recommendation ITU-R 218/3.

2.2.2 Comparison of IRI modeled results with satellite measurements

The recent expansion of satellite missions with GPS receivers onboard for precise orbit determination and radio occultation has improved the understanding of the topside ionosphere. This has been important for continuous validation and improvement of empirical models to accurately predict TEC and electron density variations for different space weather phenomena [36, 37]. The comparison of satellite measurements with models is based on error assessment using root mean square, absolute and relative values, and other statistical parameters. Previous studies found a consistent underestimation of TEC recorded by a dual-frequency altimeter on a TOPEX satellite at an altitude of 1336 km by the IRI model mostly occurring during high solar activity, at the equatorial anomaly, and the high latitude regions [36, 38]. The significant error comes from the systematic technique of TEC measurement from the altimeter and the absence of experimental territory results over ocean surfaces for IRI validation [39]. The orbit altitude over which TOPEX TEC is approximated is 1336 km which is less than 2000 km for IRI, this is also a possible cause of discrepancy between the two TEC values.

Further inconsistencies between IRI modeled and SWARM Langmuir probe electron density values are observed by Singh et al. [40] and Pignalberi et al. [37]. SWARM is a constellation of 3 satellites each with a GPS receiver, A and C fly at 460 km altitude and B at an altitude of 540 km. The in-situ electron density and TEC measurements give a more detailed description of the topside ionosphere by simultaneous measurements at different altitudes. This has been used in modeling and improving the topside formulation for the validation of empirical models [41]. In addition, the comparison of IRI modeled values with the satellite data depend on the selected input model parameters. Arıkan, et al. [42] compared the IRI-2016 F2 Layer Model Parameters with Ionosonde Measurements. He noted that the "ON" storm model improves the performance of the IRI during severe storm times. In addition, the CCIR option for peak F2 electron density produced more accurate modeled results in the low latitudes and the southern hemisphere while the URSI option is recommended for the northern hemisphere and mid-latitudes. Also, the IRI model with the NeQuick topside improved its performance when compared with the TOPEX data [43].

2.2.3 The coupled thermosphere ionosphere plasmasphere electrodynamics model (CTIPe)

CTIPe model [44] combines the global thermosphere model [45], high latitude ionosphere model [46], low and mid-latitude ionosphere/plasmasphere model [47] and an extension of the electrodynamic circulation [48]. The model inputs include F10.7 averaged over 41 days for ionospheric, heating, and dissociation processes, high latitude electric potential patterns generated by 1 minute average from ACE satellite, solar wind density, solar wind speed, interplanetary magnetic field (IMF) in the XYZ planes, tilt, and total magnetic field provided by the Space Weather Prediction Centre (SWPC) real-time databases; these are coupled with the Weimer model to produce particle and auroral precipitation patterns [49]. The model runs 30 minutes ahead of real-time with a 10-minute interval of updated results and has been adopted at the SWPC by NOAA to study the thermosphere-ionosphere conditions for space weather nowcasting and forecasting purposes.

The model simulations have reproduced ionospheric thermosphere conditions of TEC and electron density during quiet and storm periods [50, 51]. The discrepancies observed when the simulated results are compared with experimental data result from non-representation of ionospheric complex electrodynamic processes associated with plasma transport and prompt penetration electric fields (PPEF) [51] and delayed ionospheric response to forcing parameters related to space weather conditions [50, 52]. From CTIPe simulations, Vaishnav et al. [53] revealed that the ionospheric delay to solar flux changes increases with reduced eddy diffusion. This effect is more observable in the mid and low-latitude regions due to stronger solar activity and EUV acting as the forcing parameters for ionization processes. Therefore, in addition to not reliable results and uneasy accessibility for model runs with longer processing times, the estimation of correct and accurate forcing parameters also significantly affects the CTIPe modeled electron density and TEC results.

3. Special features of ionospheric variability for space weather

Ionospheric models monitor the quite long-term ionospheric (climatological) conditions occurring repeatedly and solely dependent on the sun's activity. These represent reference ionospheric conditions of electron density and TEC of the upper atmosphere influenced by diurnal, seasonal, latitudinal, local time, and solar cycle. Ionospheric storms and associated effects disturb quite ionospheric conditions with sudden changes in electron density and TEC, that is negative and positive storms [54]. The variability of the lower ionospheric (D, E & F1) regions is unique and complex to understand since its variability is influenced by forcing from the lower atmospheric and tidal processes in addition to solar activity. The latitude location adversely affects the distribution of electron density and TEC with changes in thermospheric composition and neutral winds induced by direct solar wind coupling at the high latitudes and the equatorial anomaly at the low latitudes.

3.1 Ionospheric storms

The strength of an ionospheric storm can be defined by changes in TEC and electron density while geomagnetic indices define the intensity of the geomagnetic storm. However, there is not necessarily a strong positive correlation between TEC and geomagnetic activity indices. Positive storms are more observed during the main phase while negative storms occur regularly during recovery; this is observable at all latitudes [55]. Latitudinal location determines ionospheric storm response effects (**Figure 3b**) with the penetration of magnetospheric electric fields from high to low latitudes; this causes wind perturbations and thermospheric composition changes associated with ExB drift and plasma transport processes. The result is positive storms observed at mid and low latitudes while the negative storms at the geomagnetic equator at storm onset [54, 56].

Physics-based models have shown tremendous progress in modeling ionospheric storm response effects, the Coupled Magnetosphere Ionosphere Thermosphere Model (CMIT) simulated the electron density and TEC for three storms showing positive and negative storms and thermospheric composition changes [56], (CMIT is a physics-based model coupling Lyon-Fedder-Mobarry global magnetosphere MHD code [57] with Thermosphere-Ionosphere- Electrodynamics General Circulation Model (TIE-GCM) [48], CTIPe simulations on storm day (20th November 2003) produced storm onset features, TEC and neutral density changes [58], GGCM-CTIM produced localized effects of auroral precipitation and electric fields for strong Joule heating, thermospheric upwelling, and neutral density enhancement on storm day (August 24, 2005) [59] (GGCM is Open Global General Circulation Model which calculates, 3-D global magnetosphere and 2-D high-latitude ionosphere solving resistive MHD equations with solar wind input). A more coupled model RCM-CTIPe which considers the high latitude field-aligned currents describes the electrodynamic coupling to reproduce the behavior of storm time electric field [60], Rice Convection Model (RCM)–CTIPe Model is a self-consistency that describes the electrodynamic coupling between magnetosphere, thermosphere, ionosphere and plasmasphere

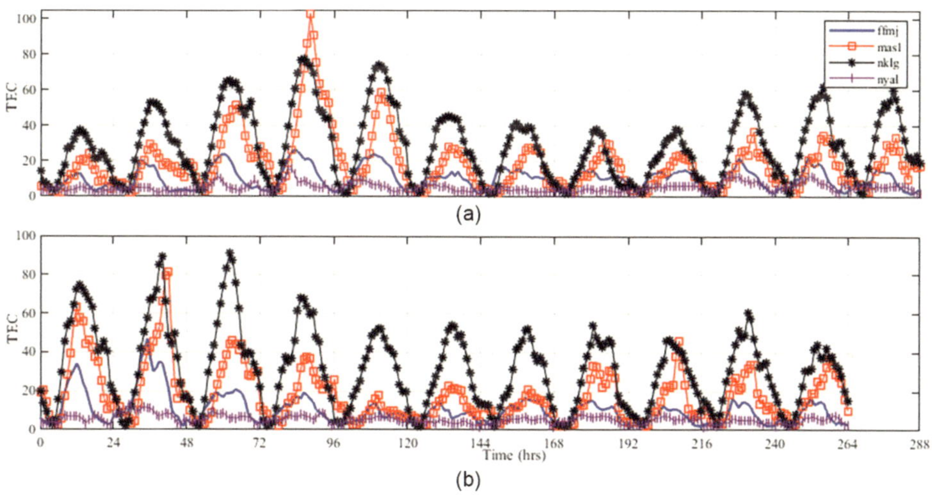

Figure 3.
Comparison of GPS-TEC at different latitudes and different storm and quiet days in 2015. Figure a and b shows the TEC variation for the 4 stations during the quietest and disturbed day of each month in 2015 respectively. The unmarked line is for ffmj (50.091°, 8.665°) mid-latitude, the square marked line is mas1 (27.764°, −15.633°) low latitude, the star marked line is for nklg (0.354°, 9.672°) the equatorial anomaly station and the line marked by 1 is nyal (78.930°, 11.865°) a high latitude station. The stations were selected within the same longitudinal band between 8° to 15° to avoid local time effects. The quietest and most disturbed days were got from the international list of quiet and disturbed days at the WDC for geomagnetism, Kyoto (http://wdc.kugi.kyoto-u.ac.jp/qddays/index.html). The RINEX files were got from CDDIS (https://cddis.nasa.gov/Data_and_Derived_Products/GNSS/daily_30second_data.html) and processed using GOPI software (https://seemala.blogspot.com/).

systems. It couples three models together, that is CTIPe [61], RCM [62], and TIE-GCM [48]. From this point of view, the self-consistency theoretical models coupling the magnetosphere, plasmasphere, ionosphere, and thermosphere systems are more dependable for reproducing storm ionospheric effects in electron density and TEC variations. However, empirical studies of storm time modeling depend on predicting the behavior of the foF2 parameter during storm periods. The storm time ionospheric correction model developed by Araujo-Pradere et al. [63] was incorporated in IRI-2000 [64]. However, the percentage deviation between experimental and IRI-2007 predicted foF2 was 100% (high) when evaluated at low latitudes [65]. Therefore, it's difficult to represent the complex dynamic processes in the data-driven models with numerical and statistical methods.

3.2 Latitudinal features

3.2.1 Electron density and TEC at high latitudes

The plasma and density redistribution at high latitudes are density depletion zones (**Figure 3**), aurora ovals, and polar cap patches. These features are due to energetic particle precipitation, open magnetic field lines, direct IMF, solar wind coupling, and induced polar cap convection [66]. TDIM, a physics-based mid-latitude model [67] was modified to include high latitude ionospheric effects of particle precipitation, electric field convection, plasma transport processes, electron density, and electron temperatures [68]. The IRAP plasmasphere-ionosphere model (IPIM), a 3-D multi-fluid mathematical model for interhemispheric studies [69] showed that electron density depletion results from a decrease of neutral atomic oxygen concentration and O/N_2 ratio; these results also closely match with the ionosonde and EISCAT radar data [70]. Theoretical models generally use statistic inputs that define precipitation, convection, and conductance rates to reproduce the complex mechanisms occurring in the auroral regions. The recommendation by the COSPAR Scientific Assembly [71] was implemented in 2013 where auroral boundaries and storm time model for aurora regions were integrated into the IRI model [72]. This has improved the performance of the IRI model in predicting the electron density and TEC over the high latitude regions [73].

3.2.2 Low latitude effects on TEC and electron density

The main features of interest in the low latitudes are the equatorial ionization anomaly (EIA) and the equatorial plasma fountain (EPF), these impact TEC and electron density variation in the F region. A recent review [74] clears the misinterpretation regarding the development and formation of the two related low latitude phenomena. The EIA removes plasma from the magnetic equator by the upward ExB drift to form a trough at the magnetic equator and ionization crests on either side of magnetic latitudes about ±15°-30°; this varies from quiet to disturbed conditions (**Figure 3**). This phenomenon is controlled by the equatorial electrojet (EEJ) which causes the fountain effect during quiet times. The EPF determines the resultant direction of plasma flow because of field-aligned plasma diffusion and field perpendicular ExB plasma drift. During disturbed days, the EIA becomes strong because of storm time equatorward wind and eastward (westward) PPEF during the main (recovery) phase associated with O/N_2 composition changes. Using multi-instrument and long-term monitoring observations, the strength and asymmetry of the anomaly, time of formation, and crest to trough ratio parameters display seasonal and diurnal variability from quiet to active solar activity and geomagnetic storms [75, 76] and references therein. Theoretical models are superior in representing the equatorial anomaly features; SUPIM [77] incorporates *ExB* drift and neutral wind velocities to produce the plasma fountain (reverse effect) during upward drift (downward drift) and through this, the EIA structures have been modeled during both quiet and active solar periods [74], Bhuyan et al. [78] reproduced electron density and temperature distribution with reasonable accuracy and the EIA phenomena around the Indian equatorial and low latitude regions using theoretical simulation. Several data-driven models for the low latitude have been developed [79, 80] and their results compared with IRI and actual measurements, this is important for continuous validation and improvement of the performance of global and regional ionospheric models.

Equatorial plasma bubbles (EPB) are plasma density fluctuations occurring when the bottom side F layer is raised above the upper limit which is dependent on solar activity, the major drivers are *ExB* plasma drift caused by pre-reversal enhancement in addition to Rayleigh–Taylor instability, gravity waves, neutral winds, and electric field fluctuations [74]. Recent modeling efforts include data-driven and numerical simulation models [81, 82], these are important to relate the nonlinear occurrence of EPBs not easily studied through simulation by theoretical models. Most numerical models cannot predict the day-to-day variability of EPBs; therefore, a need to add real-time assimilation methods of data and integrate the combined models into the global ionosphere and thermosphere electron density models for the prediction of such phenomena.

3.3 Plasma irregularities at low altitudes

3.3.1 Traveling ionospheric disturbances (TIDs)

The dynamic change in thermal neutral density within the mesosphere-lower thermosphere (MLT) altitudes depends strongly on low atmospheric forcing associated with internal gravity wave propagation rather than solar and geomagnetic activity [83, 84]. TIDs are simply gravity waves (GWs) signatures resulting from the interaction of GWs with ionospheric plasma traveling at ionospheric altitudes. Traveling atmospheric disturbances (TADs) also manifest with modification of neutral and electron density in the low altitude ionosphere. These are initiated by fluctuating aurora electrojet and or particle precipitation from enhanced heating during storm periods and their intensity varies with the solar cycle and geomagnetic activity [84]. The traveling disturbances exhibit a very complex nature of waves mixed with plasma instabilities associated with propagation characteristics and corresponding electron densities which requires classification based on wavelengths, magnitude, direction, seasonal and diurnal factors. The long-time monitoring and observations by ground and space-based instruments have supported the development and validation of empirical models [85] and semi-empirical models [86, 87] to predict spatial and temporal dynamics of TID parameters at the MLT. However, much work needs to be done regarding the associated density variations to integrate them into the global TEC models for space weather forecasting.

3.3.2 Sporadic E layers (Es)

The irregular and dynamic layers of compressed and enhanced ionization form below the F region and descend through the E-F valley region as observed by ionosonde and ISR. They are associated with electron density fluctuations generated from the combined effects of sheer wind mechanisms and electric field structures of the earth's magnetic field. While there is no significant correlation with geomagnetic activity, it's important to statistically understand their occurrence mechanisms and origin to formulate correct inputs for empirical prediction models. The Ground-to-topside model of Atmosphere and Ionosphere for Aeronomy (GAIA) has been used for numerical simulation and prediction of Es events using the horizontal wind model (HWM14) [88]. In addition to more recent efforts for numerical and data-driven models, the statistic trend of the critical frequency of Es layer (foEs) has been studied; this has shown diurnal, seasonal, geographical dependence with spatial and temporal variations [88, 89] and references therein.

4. Deep learning and neural networks

Artificial Neural Networks (ANN) have become dependable methods for modeling ionospheric variability to predict space weather and associated effects. This is because they can solve the high and complex nonlinearity involved in the magnetosphere-ionosphere-thermosphere coupling processes without the knowledge of physical, chemical, and coupling processes and be used for forecasting. The ANNs are data-driven models with massive parallel distributed processing units (neurons) orderly arranged in layers that learn nonlinear patterns (inputs) aided by a learning algorithm with logical adjustment of weights and bias. The input parameters are trained with varying numbers of hidden layers and neurons, training algorithms, and training functions till they have generalized well to a required output as TEC or electron density, refer to [90, 91] for details in computation techniques in neural networks.

Generally, the input parameters are selected based on their effect on the ionospheric variation, that is spatial and temporal variation represented by geographical positions, date and time, geomagnetic activity represented by Dst, Kp, Ap, and SYM/H indices, solar activity represented by F10.7 index, sunspot number, and proton flux and EUV flux. While CMEs and solar flares enhance the geomagnetic activity causing geomagnetic storms, the intensity of the storm activity depends on the state of the interplanetary field described by solar wind parameters [92]. Several studies have also revealed that there is a positive correlation between TEC and solar wind parameters mostly during storm days [93]. Therefore, other input parameters considered include IMF in the XYZ plane, total magnetic field, solar wind speed, temperature, density, and electric field. The specific inputs are selected for specific NN models based on the purpose and or type of the model (regional or single station model, long time or short time prediction model, storm time models). Recently, the addition of IRI's parameters related to TEC and electron density variation like peak density (NmF2), critical frequency (foF2), and peak height (hmF2) at F2 has proved to improve the capacity of NN models to learn long term trends of solar cycle variations [94, 95].

Neural networks have been applied for other special cases of modeling for example prediction of topside electron content using SWARM data [96], topside ionosphere, and plasmasphere model for electron density and TEC [97], bottom side electron density profile model over Grahamstown, South Africa [98]. With the progressive development of computation techniques and the availability of extensive datasets for space weather and ionospheric parameters, researchers are now able to develop more reliable and accurate NN models that can capture most spatial and temporal TEC and electron density variations. And for this reason, NN models are flexible and have become reference points to validate and improve the existing global and regional empirical models. However, the development of such comprehensive NN models requires very large datasets which may not be available in certain areas and at certain times mostly in African equatorial regions, strong knowledge of computational techniques, and long times for computational processing.

5. Summary and conclusion

We reviewed the development of ionospheric models and discussed several improvements and updates made. Throughout the chapter, we emphasized the need for continuous validation of ionospheric models with updated datasets. Theoretical modeling has improved from 1-D to 3-D representing the M-I-T coupling, physical and chemical processes occurring during special space weather conditions. This is important in explaining and understanding the energy transfer mechanisms from the solar wind to the earth's atmosphere during storms, the effect of wind and thermospheric composition on ionospheric TEC and electron density variations, and the effect of forcing from the lower atmosphere. Empirical models have been improved to globally predict the climatological condition of TEC and electron density like IRI. The wide network of ground monitoring and observation stations with space-based measurements has provided extended databases for space weather and ionospheric parameters. This has supported the development of reliable and accurate neural network models for prediction. The role of International and national Organizations like COSPAR, URSI, ITU, CCIR, and research and education institutions has been emphasized in supporting and maintaining ground stations for ionospheric observation. These measurements have become backbones for continuous validation and improvement of ionospheric models. However, it's important to note that some regions still have a sparse and non-uniform distribution of ground monitoring stations which is a challenge for developing representative data-driven models.

Acknowledgements

"I am also grateful to E-JUST-TICAD 7 scholarship, a partnership program between the government of Japan and the Egyptian government for the opportunity". "I am grateful to Prof. Ayman Mahrous (E-JUST) and Dr. Ola A. Abuelezz (Helwan University) for their continuous guidance and support".

Author details

Wellen Rukundo
Space Environment and Research Laboratory, Institute of Basic and Applied Science, Egypt Japan University of Science and Technology (E-JUST), Alexandria, Egypt

*Address all correspondence to: wellen.rukundo@ejust.edu.eg

References

[1] Yaacob N, Abdullah M, Ismail M. GPS total electron content (TEC) prediction at ionosphere layer over the equatorial region. In: Trends in Telecommunications Technologies. Rijeka, Croatia: InTech; 2010. pp. 485-508

[2] Yizengaw E, Moldwin MB, Galvan D, Iijima BA, Komjathy A, Mannucci AJ. Global plasmaspheric TEC and its relative contribution to GPS TEC. Journal of Atmospheric and Solar-Terrestrial Physics. 2008;**70**(11-12):1541-1548

[3] Wang C, Xin S, Liu X, Shi C, Fan L. Prediction of global ionospheric VTEC maps using an adaptive autoregressive model. Earth, Planets and Space. 2018;**70**(1):14

[4] Orús R, Hernández-Pajares M, Juan JM, Sanz J. Improvement of global ionospheric VTEC maps by using kriging interpolation technique. Journal of Atmospheric and Solar-Terrestrial Physics. 2005;**67**(16):1598-1609

[5] Chakraborty M, Kumar S, De BK, Guha A. Latitudinal characteristics of GPS derived ionospheric TEC: A comparative study with IRI 2012 model. Annals of Geophysics. 2014;**57**(5):13

[6] Bilitza D, Reinisch BW. International reference ionosphere 2007: Improvements and new parameters. Advances in Space Research. 2008; **42**(4):599-609

[7] Radicella SM, Zhang ML. The improved DGR analytical model of electron density height profile and total electron content in the ionosphere. Annals of Geophysics. 1995;**38**(1):7

[8] Leitinger R, Radicella S, Hochegger G, Nava B. Diffusive equilibrium models for the height region above the F2 peak. Advances in Space Research. 2002;**29**(6):809-814

[9] Nava B, Coïsson P, Radicella SM. A new version of the NeQuick ionosphere electron density model. Journal of Atmospheric and Solar-Terrestrial Physics. 2008;**70**:1856-1862

[10] European Union, "European GNSS (Galileo) Open Service Signal in Space Interface Control Document, OS SIS ICD, Iss 2.0," European Union, European, 2021

[11] Liu A, Wang N, Li Z, Wang Z, Yuan H. Assessment of NeQuick and IRI-2016 models during different geomagnetic activities in global scale: Comparison with GPS-TEC, dSTEC, Jason-TEC and GIM. Advances in Space Research. 2019;**63**(12):3978-3992

[12] Cherniak I, Zakharenkova I. NeQuick and IRI-Plas model performance on topside electron content representation: Spaceborne GPS measurements. Radio Science. 2016;**51**(6):752-766

[13] Tariku YA. Comparison of performance of the IRI 2016, IRI-Plas 2017, and NeQuick 2 models during different solar activity (2013-2018) years over South American sector. Radio Science. 2020;**55**(8):17

[14] Hedin AE. MSIS-86 Thermospheric Model. Journal of Geophysical Research: Space Physics. 1987;**92**(A5):4649-4662

[15] Hedin AE, Fleming EL, Manson AH, Schmidlin FJ, Avery SK, Clark RR, et al. Empirical wind model for the upper, middle and lower atmosphere. Journal of Atmospheric and Terrestrial Physics. 1996;**58**(13):1421-1447

[16] Daniell RE, Brown LD, Anderson DN, Fox MW, Doherty PH, Decker DT, et al. Parameterized ionospheric model: A global ionospheric parameterization based on first principles models. Radio Science. 1995;**30**(5):1499-1510

[17] Teters LR, Lloyd JL, Haydon GW, Lucas DL. Estimating the Performance of Telecommunication Systems Using the Ionospheric Transmission Channel. Ionospheric Communications Analysis and Prediction Program User's Manual. U.S.A.: NASA STI/Recon Technical Report N, Boulder; 1983

[18] Llewellyn S, Bent R. Documentation and description of the Bent ionospheric model. Tech. Rep. AFCRL-TR-73-0657. Massachusetts: Air Force Cambridge Res. Lab., Hansom Air Force Base; 1973. [Online]. Available from: http://oai.dtic.mil/oai/oai?verb=getRecord&metadataPrefix=html&identifier=AD0772733

[19] Barghausen AF, Finney JW, Proctor LL, Schultz LD. Predicting Long-Term Operational Parameters of High-Frequency Sky-Wave Telecommunication Systems. Boulder, Colorado: U.S. Department of Commerce, Environmental Sciences Services Administration, Research Laboratories, Institute for Telecommunication Sciences; 1969

[20] Bradley PA, Dudeney JR. A simple model of the vertical distribution of electron concentration in the ionosphere. Journal of Atmospheric and Terrestrial Physics. 1973;**35**(12):2131-2146

[21] Dudeney JR. An improved model of the variation of electron concentration with height in the ionosphere. Journal of Atmospheric and Terrestrial Physics. 1978;**40**:195-203

[22] Stankov SM, Jakowski N, Heise S, Muhtarov P, Kutiev I, Warnant R. A new method for reconstruction of the vertical electron density distribution in the upper ionosphere and plasmasphere. Journal of Geophysical Research: Space Physics. 2003;**108**(A5):21

[23] Rush CM, Miller D. A Three-Dimensional Ionospheric Model Using Observed Ionospheric Parameters. Bedford, Massachusetts: Air Force Cambridge Research Laboratories; 1973

[24] Nisbet JS, Divany R. PC version of ionospheric model available. American Geophysical Union. 1987;**68**(34):714-714

[25] Bilitza D, Altadill D, Truhlik V, Shubin V, Galkin I, Reinisch B, et al. International Reference Ionosphere 2016: From ionospheric climate to real-time weather predictions. Space Weather. 2017;**15**(2):418-429

[26] Hochegger G, Leitinger R. Inversions of satellite to satellite electron content: Simulation studies with NeUoG-plas. Physics and Chemistry of the Earth, Part C: Solar, Terrestrial & Planetary Science. 2000;**25**(4):353-357

[27] Chiu YT. An improved phenomenological model of ionospheric density. Journal of Atmospheric and Terrestrial Physics. 1975;**37**:1563-1570

[28] Köhnlein W. A model of the terrestrial ionosphere in the altitude interval 50-4000 km II. Molecular ions (N2 +, NO+, O2 +) and electron density. Earth, Moon, and Planets. 1989;**7**:109-163

[29] Anderson DN, Mendillo M, Herniter B. A semi-empirical low-latitude ionospheric model. Radio Science. 1987;**22**(2):292-306

[30] Tascione TF, Kroehl HW, Creiger R, Freeman JW, Wolf RA, Spiro RW, et al.

New ionospheric and magnetospheric specification models. Radio Science. 1988;**23**(3):211-222

[31] Reilly MH, Singh M. Ionospheric specification from GPS data and the RIBG ionospheric propagation model. Radio Science. 1997;**32**(4):1671-1679

[32] Gulyaeva TL, Bilitza D. Towards ISO Standard Earth Ionosphere and Plasmasphere Model./New Developments in the Standard Model. Vol. 39. Amsterdam: Elsevier; 2012. pp. 1-39

[33] Alcay S, Oztan G. Analysis of global TEC prediction performance of IRI-PLAS model. Advances in Space Research. 2019;**63**(10):3200-3212

[34] Adebiyi SJ, Ikubanni SO, Adebesin BO, Adeniyi JO, Joshua BW, Adimula IA, et al. Storm time IRI-Plas model forecast for an African equatorial station. Heliyon. 2019;**5**(6):8

[35] Amaechi PO, Oyeyemi EO, Akala AO, Kaab M, Younas W, Benkhaldoun Z, et al. Comparison of ionospheric anomalies over African equatorial/low-latitude region with IRI-2016 model predictions during the maximum phase of solar cycle 24. Advances in Space Research. 2021;**68**(3):1473-1484

[36] Migoya Orué YO, Radicella SM, Coïsson P, Coïsson P, Ezquer RG, Nava B. Comparing TOPEX TEC measurements with IRI predictions. Advances in Space Research. 2008;**42**(4):757-762

[37] Pignalberi A, Pezzopane M, Tozzi R, De Michelis P, Coco I. Comparison between IRI and preliminary Swarm Langmuir probe measurements during the St. Patrick storm period. Earth, Planets and Space. 2016;**68**(93):18

[38] Yasukevich YV, Afraimovich EL, Palamarchuk KS, Tatarinov PV. Testing of the international reference ionosphere model using the data of dual-frequency satellite altimeters "Topex"/"Poseidon" and "Jason-1". Radiophysics and Quantum Electronics. 2009;**52**(341):341-353

[39] Astafyeva EI, Afraimovich EL, Oinats AV, Yasukevich YV, Zhivetiev IV. Dynamics of global electron content in 1998-2005 derived from global GPS data and IRI modeling. Advances in Space Research. 2008;**42**(4):763-769

[40] Singh AK, Maltseva O, Panda SK. Comparison between Swarm measured and IRI-2016, IRI-Plas 2017 modeled electron density over low and mid latitude region. Acta Astronautica. 2021;**189**:476-482

[41] Pezzopane M, Pignalberi A. The ESA Swarm mission to help ionospheric modeling: A new NeQuick topside formulation for mid-latitude regions. Scientific Reports. 2019;**9**(12253):12

[42] Arıkan F, Sezen U, Gulyaeva TL. Comparison of IRI-2016F2 layer model parameters with ionosonde measurements. Journal of Geophysical Research: Space Physics. 2019;**124**(10):8092-8109

[43] Coïsson P, Radicella SM, Leitinger R, Ciraolo L. Are models predicting a realistic picture of vertical total electron content? Radio Science. 2004;**39**:7

[44] Millward GH, Müller-Wodarg ICF, Aylward AD, Fuller-Rowell TJ, Richmond AD, Moffett RJ. An investigation into the influence of tidal forcing on Fregion equatorial vertical ion drift using a global ionosphere-thermosphere model with coupled electrodynamics. Journal of Geophysical Research. 2001;**106**(A11):24733-24744

[45] Fuller-Rowell TJ, Rees D. A three-dimensional time-dependent

global model of the thermosphere. Journal of the Atmospheric Sciences. 1980;**37**(11):2545-2567

[46] Quegan S, Bailey GJ, Moffett RJ, Heelis RA, Fuller-Rowell TJ, Rees D, et al. A theoretical study of the distribution of ionization in the high-latitude ionosphere and the plasmasphere: First results on the mid-latitude trough and the light-ion trough. Journal of Atmospheric and Terrestrial Physics. 1982;**44**(7):619-640

[47] Bailey GJ. The effect of a meridional E×B drift on the thermal plasma at L = 1.4. Planetary and Space Science. 1983;**31**(4):389-409

[48] Richmond AD, Ridley EC, Roble RG. A thermosphere/ionosphere general circulation model with coupled electrodynamics. Geophysical Research Letters. 1992;**19**(6):601-604

[49] Fuller-Rowell TJ, Evans DS. Height-integrated Pedersen and Hall conductivity patterns inferred from the TIROS-NOAA satellite data. Journal of Geophysical Research: Space Physics. 1987;**92**(A7):7606-7618

[50] Vaishnav R, Schmölter E, Jacobi C, Berdermann J, Codrescu M. Ionospheric response to solar extreme ultraviolet radiation variations: Comparison based on CTIPe model simulations and satellite measurements. Annales of Geophysics. 2021;**39**(2):341-355

[51] Fernandez-Gomez I, Fedrizzi M, Codrescu MV, Borries C, Fillion M, Fuller-Rowell TJ. On the difference between real-time and research simulations with CTIPe. Advances in Space Research. 2019;**64**(10):2077-2087

[52] Jacobi C, Jakowski N, Schmidtke G, Woods TN. Delayed response of the global total electron content to solar EUV variations. Advances in Radio Science. 2016;**14**:175-180

[53] Vaishnav R, Jacobi C, Berdermann J, Codrescu M, Schmölter E. Role of eddy diffusion in the delayed ionospheric response to solar flux changes. Annales of Geophysics. 2021;**39**(4):641-655

[54] Buonsanto MJ. Ionospheric Storms: A Review. Space Science Reviews. 1999;**88**(3/4):563-601

[55] Iyer S, Mahajan A. Granger causality analysis of deviation in total electron content during geomagnetic storms in the equatorial region. Journal of Engineering and Applied Science. 2021;**68**(1):1-25

[56] Wang W, Lei J, Burns AG, Solomon SC, Wiltberger M, Xu J, et al. Ionospheric response to the initial phase of geomagnetic storms: Common features. Journal of Geophysical Research. 2010;**115**(A7):18

[57] Lyon JG, Fedder JA, Mobarry CM. The Lyon–Fedder–Mobarry (LFM) global MHD magnetospheric simulation code. Journal of Atmospheric and Solar-Terrestrial Physics. 2004;**66**(15-16):1333-1350

[58] Fernandez-Gomez I, Borries C, Codrescu M. Analysis of the 20th November 2003 extreme geomagnetic storm using CTIPe model and GNSS data. In: American Geophysical Union, Fall General Assembly 2016, San Francisco, California. 2016

[59] Connor HK, Zesta E, Fedrizzi M, Shi Y, Raeder J, Codrescu MV, et al. Modeling the ionosphere-thermosphere response to a geomagnetic storm using physics-based magnetospheric energy input: OpenGGCM-CTIM results. Journal of Space Weather and Space Climate. 2016;**6**(A25):15

[60] Maruyama N, Fuller-Rowell TJ, Codrescu MV, Anderson D, Richmond AD,

Maute A, et al. Modeling the storm time electrodynamics. In: Aeronomy of the Earth's Atmosphere and Ionosphere. IAGA Special Sopron Book Series. Vol. 2. Dordrecht: Springer; 2011. pp. 455-464

[61] Millward GH, Moffett RJ, Quegan S, Fuller-Rowell TJ. The coupled thermosphere ionosphere plasmasphere. In: Schunk RW, editor. STEP Handbook on Ionospheric Models. Logan: Utah State University; 1996. pp. 239-279

[62] Toffoletto F, Sazykin S, Spiro R, Wolf R. Inner magnetospheric modeling with the Rice Convection Model. Space Science Reviews. 2003;**107**(1/2):175-196

[63] Araujo-Pradere EA, Fuller-Rowell TJ, Codrescu MV. STORM: An empirical storm-time ionospheric correction model 1. Model description. Radio Science. 2002;**37**(5):3.1-3.12

[64] Bilitza D. International reference ionosphere 2000. Radio Science. 2001;**36**(2):261-275

[65] Oyeyemi EO, Adewale AB, Adeloye AB, Olugbon B. An evaluation of the IRI-2007 storm time model at low latitude stations. Advances in Space Research. 2013;**52**(10):1737-1747

[66] Tsunoda RT. High-latitude F region irregularities: A review and synthesis. Reviews of Geophysics. 1988;**26**(4):719-760

[67] Schunk RW, Walker JCG. Theoretical ion densities in the lower ionosphere. Planetary and Space Science. 1973;**21**(11):1875-1896

[68] Schunk RW. A mathematical model of the middle and high latitude ionosphere. Pure and Applied Geophysics. 1988;**127**(2-3):255-303

[69] Marchaudon A, Blelly P-L. A new interhemispheric 16-moment model of the plasmasphere-ionosphere system: IPIM. Journal of Geophysical Research: Space Physics. 2015;**120**(7):5728-5745

[70] Marchaudon A, Blelly P-L, Grandin M, Aikio A, Kozlovsky A, Virtanen I. IPIM modeling of the ionospheric F2 layer depletion at high latitudes during a high-speed stream event. Journal of Geophysical Research: Space Physics. 2018;**123**(8):7051-7066

[71] Bilitza D. The High Latitudes in the International Reference Ionosphere (IRI). Hamburg, Germany: COSPAR Scientific Assembly; 1994

[72] Bilitza D. The International Reference Ionosphere – Status 2013. Advances in Space Research. 2015;**55**(8):1914-1927

[73] Timoçin E, Temuçin H, Inyurt S, Shah M, Jamjareegulgarn P. Assessment of improvement of the IRI model for foF2 variability over three latitudes in different hemispheres during low and high solar activities. Acta Astronautica. 2021;**180**:305-316

[74] Balan N, Liu L, Le H. A brief review of equatorial ionization anomaly and ionospheric irregularities. Earth and Planetary Physics. 2018;**2**(4):1-19

[75] Kwak Y-S, Kil H, Lee WK, Yang T-Y. Variation of the hemispheric asymmetry of the equatorial ionization anomaly with solar cycle. Journal of Astronomy and Space Sciences. 2019;**36**(3):159-168

[76] Amaechi PO, Oyeyemi EO, Akala AO. Variability of the African equatorial ionization anomaly (EIA) crests during the year 2013. Canadian Journal of Physics. 2019;**97**(2):24

[77] Bailey GJ, Balan N, Su YZ. The Sheffield University plasmasphere ionosphere model: A review. Journal

of Atmospheric and Solar-Terrestrial Physics. 1997;**59**(13):1541-1552

[78] Bhuyan PK, Borgohain A, Bhuyan K. Theoretical simulation of electron density and temperature distribution at Indian equatorial and low latitude ionosphere. Advances in Space Research. 2008;**41**(4):587-598

[79] Hajra R, Chakraborty SK, Tsurutani BT, Dasgupta A, Echer E, Brum CGM, et al. An empirical model of ionospheric total electron content (TEC) near the crest of the equatorial ionization anomaly (EIA). Journal of Space Weather and Space Climate. 2016;**6**(A29):9

[80] Feng J, Han B, Zhao Z, Wang Z. A new global total electron content empirical model. Remote Sensing. 2019;**11**(6):19

[81] Yokoyama T, Shinagawa H, Jin H. Nonlinear growth, bifurcation and pinching of equatorial plasma bubble simulated by three-dimensional high-resolution bubble model. Journal of Geophysical Research: Space Physics. 2014;**119**(12):10474-10482

[82] Hysell DL, Jafari R, Milla MA, Meriwether JW. Data-driven numerical simulations of equatorial spread F in the Peruvian sector. Journal of Geophysical Research: Space Physics. 2014;**119**(5):3815-3827

[83] Borchevkina O, Karpov I, Karpov M. Meteorological storm influence on the ionosphere parameters. Atmosphere. 2020;**11**(9):14

[84] Knàzovà PK, Lastovika J, Kouba D, Mosna Z, Podolskà K, Potuznàkovà K, et al. Ionosphere influenced from lower-lying atmospheric regions. Frontiers in Astronomy and Space Sciences. 2021;**8**:29

[85] Reinisch B, Galkin I, Belehaki A, Paznukhov V, Huang X, Altadill D, et al. Pilot ionosonde network for identification of traveling ionospheric disturbances. Radio Science. 2018;**53**(3):365-378

[86] Fedorenko YP, Tyrnov OF, Fedorenko VN, Dorohov VL. Model of traveling ionospheric disturbances. Journal of Space Weather and Space Climate. 2013;**3**(A30):28

[87] Krall J, Huba JD, Ossakow SL, Joyce G, Makela JJ, Miller ES, et al. Modeling of equatorial plasma bubbles triggered by non-equatorial traveling ionospheric disturbances. Geophysical Research Letters. 2011;**38**(8):5

[88] Shinagawa H, Tao C, Jin H, Miyoshi Y, Fujiwara H. Numerical prediction of sporadic E layer occurrence using GAIA. Earth, Planets and Space. 2021;**73**(28):18

[89] Carrasco AJ, Batista IS, Abdu MA. Simulation of the sporadic E layer response to prereversal associated evening vertical electric field enhancement near dip equator. Journal of Geophysical Research: Space Physics. 2007;**112**(A6):10

[90] Haykin S. Neural Networks and Learning Machines. 3rd ed. New Jersey: Pearson Education, Inc; 2008

[91] Demuth H, Mark B. Neural Network Toolbox User's Guide, Version. 4th ed. Natick, Massachusetts: The MathWorks, Inc; 2004

[92] Subedi A, Adhikari B, Mishra RK. Variation of solar wind parameters during intense geomagnetic storms. Himalayan Physics. 2017;**6**:80-85

[93] Mishra RK, Adhikari B, Chapagain NP, Baral R, Das PK, Klausner V, et al. Variation on solar wind parameters and total electron content

over middle- to low-latitude regions during intense geomagnetic storms. Radio Science. 2020;**55**(11):26

[94] Sahu S, Trivedi R, Choudhary RK, Jain A, Jain S. Prediction of total electron content (TEC) using neural network over anomaly crest region Bhopal. Advances in Space Research. 2021;**68**(7):2919-2929

[95] Tebabal A, Radicella SM, Damtie B, Migoya-Orue Y, Nigussie M, Nava B. Feed forward neural network based ionospheric model for the East African region. Journal of Atmospheric and Solar-Terrestrial Physics. 2019;**191**:10

[96] Abuelezz OA, Mahrous AM, Cilliers PJ, Fleury R, Youssef M, Nedal M, et al. Neural network prediction of the topside electron content over the Euro-African sector derived from Swarm-A measurements. Advances in Space Research. 2021;**67**(4):1191-1209

[97] Watanabe S, Miyoshi Y, Tsuchiya F, Kumamoto A, Kasahara Y, Matsuoka A et al., "Modeling of Topside Ionosphere and Plasmasphere", 2021. p. 18

[98] McKinnell LA. A Neural Network Based Ionospheric Model for the Bottomside Electron Density Profile over Grahamstown, South Africa. Grahamstown, South Africa: Rhodes University; 2003

Chapter 6

Application of Onsager and Prigozhin Variational Principles of Nonequilibrium Thermodynamics to Obtain MHD-Equation Dissipative System in Drift Approximation

Vadim Bogdanov

Abstract

Electromagnetic phenomena in plasma are easier to describe in terms of fields, expressing the electric current through the rotor of the magnetic field. But the approach that ignores the corpuscular aspect of the electric current, as noted by H. Alfven, does not allow describing many processes in space plasma. Indeed, relying on the concept of continuity, it is impossible in the mechanics of continuous media to take into account the fluctuations of hydrodynamic functions formed due to the molecular structure of the medium. At the hydrodynamic level of description, taking into account the structure leads to the Langevin equation. Therefore, to describe processes in a magnetized plasma, it is of certain interest to obtain MHD equations in the drift approximation not from the Vlasov equations, but based on the principles of Onsager and Prigogine, combined by Gyarmati into one variational principle and obtaining a one-liquid plasma model in the drift approximation. Fluctuations are taken into account by introducing an additional term in the expression for pressure, written in the drift approximation, which is similar to the postulation of the Langevin source for describing Brownian motion. The obtained fluctuating-dissipative system differs from the reversible one-liquid approximation of the two adiabatic invariants of Chu, Goldberger, Low.

Keywords: nonequilibrium thermodynamics, variational principles of Onsager and Prigogine, the combined Gyarmati principle, collisionless plasma, drift approximation

1. Introduction

Alfven in his work [1] noted that the approach which does not take into account the corpuscular aspect of the electric current does not allow to fully describe many processes in the cosmic plasma. Relying on the concept of continuity, it is impossible in continuum mechanics to take into account fluctuations of hydrodynamic functions formed due to the molecular structure of the medium. It is known that at the hydrodynamic level of description, taking into account the corpuscular structure leads to the Langevin equation, in which the parameters of the medium are described by random sources [2]. These sources are responsible for fluctuations of density, velocity, temperature and, being the unavoidable properties of the medium, cannot be excluded. In turn, the model of "collisionless" plasma based on the Vlasov equations, in principle, does not contain fluctuations, since it is collisions that lead to fluctuations and, as a consequence, to dissipation. Naturally, magnetohydrodynamic equations (MHD equations) obtained in the drift approximation from the Vlasov equation through the moments of the distribution function also do not take into account dissipative processes (see, for example, [3]). In a magnetized plasma, the distribution of electrons and ions can have axial symmetry with respect to the magnetic field. In the absence of heat flux along the magnetic field lines (or it can be neglected), slow plasma motions obey MHD equations with anisotropic pressure. In a number of interesting cases, the description of the plasma behavior without collisions in the hydrodynamic approximation can be used as a heuristic tool for obtaining qualitatively correct results [3]. It should be noted that a significant part of the work on the macroscopic description of plasma behavior is devoted to clarifying the question of how much a real plasma can differ from its ideal twin under the assumption, for example, of an ideally conducting liquid [4].

Therefore the problem arises to try to obtain the MHD equations not from the Vlasov equations, but on the basis of another approach, in which the drift equations themselves, in the conclusion of which the perturbation theory lies [1], are the initial ones. Such a possibility opens in the case of application of the principles of the least dissipation of energy of Onsager [5] and the least production of entropy of Prigozhin [6], combined by Gyarmati into one variational principle [7]. In this case, the fluxes corresponding to the observed transport processes in a magnetized plasma are represented in the drift approximation. In turn, the drift approximation, being one-particle, simultaneously admits fluctuations within the accuracy of this approximation $T_L/H|dH| < < 1$, where T_L is the period of the Larmorian rotation.

The application of variational principles allows one to obtain a hydrodynamic system of equations, which in the linear approximation describes in the drift approximation the dynamics of a collisionless plasma located near the equilibrium state. Unlike the Vlasov equation and the equations of hydrodynamics that follow from it (or postulated on the basis of known conservation laws), the resulting system of equations is completely self-consistent and takes into account the fluctuation interaction of local currents with electric and magnetic fields within the accuracy of the used drift approximation. Fluctuations are taken into account by introducing an additional term in the expression for the pressure, which is responsible for its nonequilibrium part, which is analogous to the postulation of a Langevin source in describing Brownian particles in hydrodynamics.

2. Statement of the problem

Difficulties that arose from the very beginning after obtaining kinetic equations and introducing the terms collisional and collisionless plasma [8, 9] are associated in the physics of open systems with the concept of continuous medium. In this case, it becomes important to determine the physically infinitesimal scale corresponding to the point of the "continuous medium".

Indeed, the concept of continuous medium, depending on the chosen model for describing the behavior of an ionized gas (kinetic, diffusion, or hydrodynamic), implies the choice of a scale characterizing a physically infinitesimal point of a continuous medium ℓ_f for which differential equations are written. However, in this case, information is lost inside these points, since the large number of particles filling their volume ($g^{-1} = n\lambda_D^3 > > 1, g$ is the plasma parameter, λ_D is Debye radius) is not taken into account, which ultimately determines the internal openness of the chosen level of description [2]. Therefore, taking into account the structure due to the "artificial" introduction of an additional collision integral into the dissipative Vlasov equation when calculating the Landau collisionless damping coefficient leads to the appearance of dissipation and, as a consequence, to nonequilibrium. The need to take into account the structure of a physical "point" is one of the main provisions that determine the substantive part of fundamental works [2, 10, 11]. This position sets the direction of the search for the possibility of describing nonequilibrium processes on the kinetic and hydrodynamic scales from a single point of view, and will be used in this work.

It is known that the description of the dynamics of an ionized gas is also possible at the hydrodynamic level. Indeed, the kinetic method for some practical problems may turn out to be too detailed and mathematically complex. At the same time, without being interested in the motion and interaction of a large number of particles, one can significantly simplify the problem associated with the study of collective processes occurring in a plasma. Considering such macroscopic quantities as the average velocity of motion of a medium $\overleftarrow{V}$, pressure P, density of particles n and currents $\overleftarrow{j}$, and so on, postulating then the basic equations of hydrodynamics of continuous media, based on the laws of conservation of mass, momentum, energy and charge, together with Maxwell's equations, we can reduce the problem to the problems of magnetohydrodynamics (MHD). The system of MHD equations has the simplest form in the case of a one-fluid approximation for scalar (see, for example, [12, 13]) or tensor pressure (quasi-hydrodynamic approximation of Chu, Goldberger, and Lowe (ChGL) [14]).

At the same time, the Lorentz force acting on charged particles in a magnetic field twists them around the lines of force, preventing movement across the lines of force, and in this regard, the action of the field is similar to the effect of collisions, limiting the movement of the particle by the value of the Larmor radius. Consequently, the drift approximation shows how, in the absence of collisions, the order inherent in "collisional" continuous media and practically sufficient for describing the dynamics of a "collisionless" plasma at the hydrodynamic level is provided by a magnetic field. ("Practical sufficiency", from the point of view of the kinetic description, is achieved by neglecting the third moments in the equations, which corresponds to the not entirely justified neglect of the heat flux along the lines of force. Experimentally, this is realized in closed axial plasma systems or under real conditions, for example, in the region capture of the Earth's magnetosphere). Consequently, in a magnetized plasma, the role of the mean free path is played by the Larmor radius of ions $\rho_{L_i}\left(\rho_{L_i} > > \rho_{L_e}\right)$, and the condition for the applicability of the continuous medium approximation takes the form $L > > \rho_{L_i}$, where L is the characteristic size in the plasma. As for the frequency dependence, which makes it possible to consider a collisionless plasma as a continuous medium during the propagation of a wave process in it, it has the form: $\omega < < \omega_{L_i} < < \omega_{0_i}$ for a not too discharged ionized gas and a weak magnetic field (hot plasma) and $\omega < < \omega_{0_i} < < \omega_{L_i}$ for a magnetized plasma satisfying the drift approximation (cold plasma). Moreover, the possibility of describing the behavior of a collisionless plasma using a pressure gradient is associated with the mechanism of pressure transfer not through collisions, but through the interaction of currents flowing in the plasma drift currents and magnetizing currents. In addition, a large role in the processes occurring in a collisionless plasma is played by self-consistent fields that bind particles and prevent them from scattering.

For physically small linear and time scales ℓ_f and τ_f, as well as the number of particles N_f in the volume ℓ_f^3, the inequalities are valid $\tau_f \sim (\lambda_D/V_T) << T$, $\ell_f \sim \lambda_D$ [8]. The first inequality makes it possible to use the "continuous medium" approximation, the second - to use the concept of "collisionless plasma", and the third notes the fact that the interaction of charged particles in an ionized medium has a collective character (V_T is the thermal velocity of particles, T is the characteristic time).

However, magnetohydrodynamic equations (MHD equations) obtained in the drift approximation from the Vlasov equation through the moments of the distribution function do not take into account dissipative processes [3]. In other words, in this case, the structure of the physically small volume of the continuous medium is not taken into account, with respect to which the macroscopic equations are written. At the same time, the possibility of taking into account the drift approximation in the hydrodynamic consideration of the theory of magnetized plasma without any additional assumptions appears in the case of applying the variational principles of nonequilibrium thermodynamics of Prigogine and Onsager [5, 6], combined by Gyarmati [7]. Thus, in the mechanics of continuous media, it becomes possible to construct non-equilibrium models that describe the dynamics of continuous systems located near equilibrium (linear approximation). In turn, the construction of new models is an important section of continuum mechanics, and they are based on the search for additional relationships between the parameters that describe the state of the considered continuous medium.

With this in mind, the following provisions were the starting points for constructing a hydrodynamic model based on variational principles and drift equations [3, 7, 15]:

1. Incomplete description of plasma in the language of fields, considered as a continuous medium, which arises when currents are replaced according to Maxwell's equations by a magnetic field [1]. This leads to neglect of the corpuscular aspect of currents and, as a consequence, neglect of the fluctuation interaction, which is formed precisely due to the molecular structure of the medium. In turn, taking into account the molecular structure of a continuous medium inevitably leads to the appearance of dissipation in it.

2. The variational principle of Gyarmati [7], which combines the principles of Onsager and Prigogine [5, 6], makes it possible, within the framework of the Lagrangian formalism, to obtain the equation of motion with allowance for dissipation for a magnetized plasma (the pressure is anisotropic) in the approximation in which thermodynamic forces Xi and fluxes Ji. In this chapter, we use the drift approximation [3, 7, 15] and the approximation of two adiabatic invariants [14]. Since the ChGL approximation is holonomic, i.e. all quantities can be expressed using the displacement vector and described by the Lagrange formalism, then the dissipative approximation of the ChGl will be obtained within the framework of this formalism.

3. Equation of collisionless plasma motion considering dissipation. Anisotropic case

To describe nonequilibrium thermodynamic processes in continuous media in a linear approximation, the Hungarian physicist Gyarmati formulated a variational principle that combines the principle of the least dissipation of Onsager's energy and the principle of the least production of Prigogine's entropy. To obtain the equation of motion that takes into account dissipation, we introduce the entropy production function $\sigma = \sum_{i=1}^{n} J_i X_i$, as well as the scattering potentials $\Psi = \frac{1}{2}\sum_{i,k}^{n} L_{i,k} X_i X_k$ and $\Phi = \frac{1}{2}\sum_{i,k}^{n} R_{i,k} J_i J_k$, expressed in terms of thermodynamic forces X_i (gradients of temperature, pressure, potential, field strength, and so on), and fluxes J_i corresponding to the observed transfer processes. If we now construct a function, $L = \Psi + \Phi - \sigma$, then, as shown in [7], thermodynamic nonequilibrium processes near a steady state develop in such a way that the integral of over the volume occupied by the medium under study is minimal

$$\int_{\mho} L d\mho = \int_{\mho} [\Psi + \Phi - \sigma] d\mho = min$$

In this formulation, the Gyarmati principle is similar to Hamilton's principle in mechanics and the variation of this integral is equal to zero. Following the general provisions of [7, 11], we represent the tensor pressure of positively charged particles of an ionized gas as a sum of two parts. One part $\overset{\leftrightarrow}{P}$ depends on the state and corresponds to the equilibrium part, the other part $\overset{\leftrightarrow}{P}_d$ depends on the rate of change of this state and corresponds to the nonequilibrium part, that is

$$\overset{\leftrightarrow}{P}{}^{i}_{\Sigma} = \overset{\leftrightarrow}{P}{}^{i} + \overset{\leftrightarrow}{P}{}^{i}_{d} \tag{1}$$

the subscript "i" denotes the ionic component of the equilibrium and nonequilibrium parts of the plasma pressure tensor. From the general provisions on the form of the explicit dependence of pressure $\overset{\leftrightarrow}{P}{}^{i}_{d}$, it follows that it should depend on the macroscopic velocity of the medium $\overset{\leftarrow}{V}_i$ and on the physical reasons causing the appearance of the nonequilibrium part of the pressure (for example, for viscous media with Brownian particles, this is taken into account by introducing the corresponding coefficients of viscosity and a random Langevin source). In our case, viscosity in the usual sense is absent, and the nonequilibrium part of the equation should be proportional to the flows of charged particles, which also corresponds to the general concept of pressure transfer through electromagnetic interaction, and also takes into account the discreteness of the ionized medium (its atomic-molecular structure [2]). With this approach, the ionic component of the pressure tensor $\overset{\leftrightarrow}{P}{}^{i}_{d}$ is similar to a Langevin source. According to what has been said, we represent the nonequilibrium part of the pressure in the form

$$\overset{\leftrightarrow}{P}{}^{i}_{d} = -m_i \left(\overset{\leftarrow}{V}{}^{i}_{k} \cdot \overset{\leftarrow}{J}{}^{i}_{n} \right) \overset{\leftrightarrow}{I} \tag{2}$$

where $\overset{\leftrightarrow}{I}$ is the unit tensor, and for the equilibrium part we write out the standard representation of this part of the pressure [12]

$$\left(\overset{\leftrightarrow i}{P}\right)_{kn} = p^i_{II} \overset{\leftarrow}{e}_k \overset{\leftarrow}{e}_n + p^i_{\perp}\left(\delta_{kn} - \overset{\leftarrow}{e}_k \overset{\leftarrow}{e}_n\right), \overset{\leftarrow}{e}_1 = \frac{\overset{\leftrightarrow}{H}}{H}. \tag{3}$$

Spatial heterogeneity and concentration n are taken into account in the explicit form of the flow $\overset{\leftarrow}{J}_i$.

We represent the Gyarmati principle in the form [13],

$$\delta \int_{\mho} (\sigma_d - \Psi_d) d\mho = 0 \tag{4}$$

where $\sigma_d = \sum_{j=1}^{f} J_j X_j$ and $\Psi_d = \frac{1}{2}\sum_{j,k=1}^{f} L_{jk} X_j X_k$. The integral in (4) is taken over the entire volume $\mho$ occupied by the plasma. Since in a collisionless plasma there are no chemical reactions and sources of death and production of particles, and the interaction of currents leads to dissipative phenomena, then according to the general principles of construction σ_d and Ψ_d [7] we have for the positive plasma component

$$\sigma^i_d = -\overset{\leftrightarrow i}{P}_d : \left(\overset{\leftarrow}{\nabla} \cdot \overset{\leftarrow i}{V}\right), \Psi^i_d = \frac{1}{2} m_i \left(\overset{\leftarrow i}{V} \cdot \overset{\leftarrow i}{J}\right)\left(\overset{\leftarrow}{\nabla} \cdot \overset{\leftarrow}{V}_i\right) = -\frac{1}{2} \overset{\leftrightarrow i}{P}_d : \left(\overset{\leftarrow}{\nabla} \cdot \overset{\leftarrow i}{V}\right).$$

Considering σ_d and Ψ_d values and on the basis of (4), we obtain

$$\delta \int_{\mho} \left(-\overset{\leftrightarrow i}{P}_d : \overset{\leftarrow}{\nabla} \cdot \overset{\leftarrow i}{V} + \frac{1}{2} \overset{\leftrightarrow i}{P}_d : \overset{\leftarrow}{\nabla} \cdot \overset{\leftarrow i}{V}\right) d\mho = -\frac{1}{2} \delta \int_{\mho} \overset{\leftrightarrow i}{P}_d : \left(\overset{\leftarrow}{\nabla} \cdot \overset{\leftarrow i}{V}\right) d\mho. \tag{5}$$

To calculate the integrand, we use the equation of balance of translational kinetic energy [7]

$$\rho_i \frac{d}{dt} \frac{\left(\overset{\leftarrow i}{V}\overset{\leftarrow i}{V}\right)}{2} + \overset{\leftarrow}{\nabla} \cdot \left(\overset{\leftrightarrow i}{P}_{\Sigma} \cdot \overset{\leftarrow i}{V}\right) = \rho_i \left(\overset{\leftarrow i}{V} \cdot \overset{\leftarrow i}{F}_{ext}\right) + \overset{\leftrightarrow i}{P}_{\Sigma} : \left(\overset{\leftarrow}{\nabla} \cdot \overset{\leftarrow i}{V}\right) \tag{6}$$

where $\overset{\leftrightarrow i}{P}_{\Sigma}$ is total pressure, determined by (1), $\overset{\leftarrow i}{F}_{ext}$ is external and internal forces per mass unit, ρ_i is ion component density. If we now express $\overset{\leftrightarrow i}{P}_d : \left(\overset{\leftarrow}{\nabla} \cdot \overset{\leftarrow i}{V}\right)$ in (5) on the basis of (6), we obtain

$$-\frac{1}{2} \delta \int_{\mho} \overset{\leftarrow i}{V} \left(\rho_i \frac{d\overset{\leftarrow i}{V}}{dt} + Div\overset{\leftrightarrow i}{P} - m_i \overset{\leftarrow i}{V} \overset{\leftarrow i}{\nabla} \cdot \overset{\leftarrow i}{J} - \rho_i \overset{\leftarrow i}{F}_{ext}\right) d\mho = -\delta \int_{\mho} L_i d\mho = 0,$$

where $L_i = \frac{1}{2} \overset{\leftarrow i}{V} \left(\rho_i \frac{d\overset{\leftarrow i}{V}}{dt} + Div\overset{\leftrightarrow i}{P} - m_i \overset{\leftarrow i}{V} \overset{\leftarrow i}{\nabla} \cdot \overset{\leftarrow i}{J} - \rho_i \overset{\leftarrow i}{F}_{ext}\right)$ is Lagrange density, which satisfy the general equation

$$\frac{\partial L}{\partial V_\beta}-\sum_{\alpha=1}^{3}\frac{\partial}{\partial X_\alpha}\frac{\partial L}{\partial\left(\partial V_\beta/\partial X_\alpha\right)}=0, \tag{7}$$

which is also valid for electronic component. Substituting the value L_i into Eq. (7) and performing differentiation, we obtain the equation of motion for the ionic component "i"

$$\rho_i\frac{d\overset{\leftarrow i}{V}}{dt}=-Div\overset{\leftrightarrow i}{P}+\rho_i\overset{\leftarrow i}{F}_{ext}+2m_i\left(\overset{\leftarrow i}{V}\cdot\overset{\leftarrow i}{\nabla}\overset{\leftarrow i}{J}\right) \tag{8}$$

where Div the operator denotes tensor divergence. Repeating the same procedure for the plasma negative component, which is near the thermodynamic equilibrium ($T_e\approx T_i$), a similar equation may be obtained for electronic e component. Adding the obtained equation for electrons to (8) and considering $\overset{\leftarrow e}{V}\approx\overset{\leftarrow i}{V}=\overset{\leftarrow}{V}$, $\overset{\leftarrow i}{F}_{ext}\approx\overset{\leftarrow e}{F}_{ext}=\overset{\leftarrow}{F}_{ext}$, and $m_e+m_i=m_i(1+m_e/m_i)\approx m_i=m$, $\rho_i=m_in_i\approx\rho$ we obtain the following equation of motion:

$$\rho\frac{d\overset{\leftarrow}{V}}{dt}=-Div\overset{\leftrightarrow i}{P}+\rho\overset{\leftarrow i}{F}_{ext}+2m\overset{\leftarrow}{V}\left[\overset{\leftarrow}{\nabla}\cdot\overset{\leftarrow i}{J}+\frac{m_e}{m_i}\frac{\partial}{\partial t}(n_e-n_i)\right] \tag{9}$$

where $\overset{\leftrightarrow}{P}=\overset{\leftrightarrow e}{P}+\overset{\leftrightarrow i}{P}=\overset{\leftrightarrow e,i}{P}_{\perp}+\overset{\leftrightarrow e,i}{P}_{II}$. In (9) $\overset{\leftarrow}{\nabla}\cdot\overset{\leftarrow i}{J}$ value is expressed through $\frac{\partial}{\partial t}(n_e-n_i)$, considering the violation of quasi-neutrality and condition $1>>m_e/m_i$.

Taking into account the structure of a physically infinitesimal element of the medium, it must be remembered that it has linear dimensions of the order of the Debye radius, within which the condition of quasineutrality, due to fluctuations, can be violated. This is of fundamental importance, since it is the fluctuations that determine the character of the development of possible instabilities in the plasma. Therefore, in the last expression, the partial derivative of the difference between the concentrations of the electronic and ionic components is multiplied by a small value $(m_e/m_i)\approx(m_e/m)$.

Since in (9) the total flux is determined through the sum of fluxes of positively charged particles $\overset{\leftarrow}{J}\approx\overset{\leftarrow i}{J}=\sum_k\overset{\leftarrow i}{J}_k$, then after simple ones associated with calculating the corresponding divergences in the drift approximation for fluxes [1, 16] (see appendix), we have

$$\overset{\leftarrow}{J}_1=\frac{cn}{eH}rot\left[\frac{p_\perp}{n}\frac{\overset{\leftarrow}{H}}{H}\right],div\overset{\leftarrow}{J}_1=0, \tag{10}$$

$$\overset{\leftarrow}{J}_2=nc\frac{\left[\overset{\leftarrow}{E},\overset{\leftarrow}{H}\right]}{H^2},div\overset{\leftarrow}{J}_2=\frac{2}{mv_\perp^2}\left(\overset{\leftarrow}{E}\overset{\leftarrow}{j}_m\right)-\frac{2}{mv_\perp^2}\left(\overset{\leftarrow}{E}\overset{\leftarrow}{j}_{gr}\right), \tag{11}$$

$$\overset{\leftarrow}{J}_3=n\frac{mcv_\perp^2}{2eH^3}\left[\overset{\leftarrow}{H},\overset{\leftarrow}{\nabla}H\right],div\overset{\leftarrow}{J}_3=\frac{2}{mev_\perp^2}\left(\overset{\leftarrow}{F}_m\overset{\leftarrow}{j}_m\right), \tag{12}$$

$$\overset{\leftarrow}{J}_4=\frac{nc}{eH^2}\left[\overset{\leftarrow}{H},\overset{\leftarrow}{\nabla}\frac{p_\perp}{n}\right],div\overset{\leftarrow}{J}_4=\frac{2}{mev_\perp^2}\left(\overset{\leftarrow}{F}_m\overset{\leftarrow}{j}_m\right)-\frac{2}{mev_\perp^2}\left(\overset{\leftarrow}{F}_m\overset{\leftarrow}{j}_{gr}\right), \tag{13}$$

$$\overleftarrow{J}_5 = n\frac{mcv_{II}^2}{eH^2R^2}\left[\overleftarrow{R},\overleftarrow{H}\right], div\overleftarrow{J}_5 = \frac{2}{mev_{\perp}^2}\left[\left(\overleftarrow{F}_c\overleftarrow{j}_m\right) + e\frac{v_{\perp}^2}{v_{II}^2}\left(\overleftarrow{E}\overleftarrow{j}_c\right) + \frac{v_{\perp}^2}{v_{II}^2}\left(\overleftarrow{F}_m\overleftarrow{j}_c\right)\right], \quad (14)$$

$$\overleftarrow{J}_6 = \frac{nc}{eH^2}\left[\overleftarrow{H}, \overleftarrow{\nabla}\left(\frac{p_{II}}{n}\right)\right], div\overleftarrow{J}_6 = -\frac{2}{mev_{\perp}^2}\left[2e\left(\overleftarrow{E}\overleftarrow{j}_m\right) + 2\left(\overleftarrow{F}_m\overleftarrow{j}_m\right) - 2\left(\overleftarrow{E}\overleftarrow{j}_{gr}\right) - 2\left(\overleftarrow{F}_m\overleftarrow{j}_c\right)\right], \quad (15)$$

$$\overleftarrow{J}_7 = nv_{II}\overleftarrow{e}_1, div\overleftarrow{J}_7 = \frac{1}{mev_{II}^2}\left(\overleftarrow{F}_M\overleftarrow{j}_{II}\right) + \frac{1}{mev_{II}^2}\left(\overleftarrow{E}\overleftarrow{j}_{II}\right). \quad (16)$$

Flows $\overleftarrow{J}_{2,3}$ arise due to electric and gradient drifts. Accounting for fluxes $\overleftarrow{J}_{4,6}$ is associated with the interdependence of magnetic pressure and plasma pressure observed in the quasi-hydrodynamic approximation, since the pressure of charged particles in the absence of collisions is transferred by currents. In addition, the fluxes $\overleftarrow{J}_{4,6}$ also take into account thermal diffusion, which is associated with the temperature gradient ($p_{\perp} \sim T_{\perp}$ and $p_{\perp} \sim T_{\perp}$). The flow $\overleftarrow{J}_5$ is associated with centrifugal forces due to the curvature of the lines of force, $\overleftarrow{J}_7$ - the flow of charged particles along the line of force. Opposite the corresponding values of the fluxes, their divergences are presented, in the derivation of which the invariance n/H and μ (first adiabatic invariant) with the accuracy of the drift approximation were taken into account and the following designations were adopted [16]:

$\overleftarrow{j}_{gr} = \frac{nc}{H^2}\mu\left[\overleftarrow{H}, \overleftarrow{\nabla H}\right]$ is gradient drift current; $\overleftarrow{j}_m = -\frac{nc}{H}\mu rot\,\overleftarrow{H}$ is magnetizing current \$;

$\overleftarrow{j}_c = \frac{mv_{II}^2}{R^2}\left[\overleftarrow{R}, \overleftarrow{H}\right]$ is centrifugal drift; $\overleftarrow{F}_m = -\mu\overleftarrow{\nabla}H$ is magnetic force;

$\overleftarrow{F}_c = \frac{mv_{II}^2}{R^2}\overleftarrow{R} = 2\frac{\varepsilon_{II}}{H}\overleftarrow{\nabla}_{\perp}H$ is a force, affecting a charged particle in inhomogeneous magnetic field (centrifugal). It is clear that in this case the divergence of the flow of particles $\overleftarrow{J}_1$ is equal to the divergence of the flow of leading centers, since in an ionized medium the motion of non-interacting particles differs from the motion of leading centers only by vortex terms, therefore $div\overleftarrow{J}_1 = 0$. In addition, in deriving (10), the change in the average kinetic energy along the magnetic field line was neglected. Let us consider the second term in square brackets of (9), associated with the violation of the quasi-stationarity condition. Fluctuational charge separation in plasma leads to the appearance of an alternating electric field, which is responsible for the onset of polarization drift, which, in turn, leads to the formation of a drift polarization current $\overleftarrow{j}_p$. The magnitude of the drift current arising from the separation of charges is proportional to the rate of change in the electric field strength. This allows us to consider it as a displacement current that occurs during the polarization of dielectrics. Having carried out the appropriate calculations, and without limiting the generality of the proposed approach, we consider a special case when an alternating electric field is perpendicular to the magnetic field, we obtain (see Appendix)

$$\frac{\partial}{\partial t}(n_e - n_i) = -\frac{4H^2}{\left(H^2 + 4\pi nmc^2\right)}\frac{\overleftarrow{F}_m\overleftarrow{j}_p}{emv_{\perp}^2}. \quad (17)$$

Substituting divergence values (9), calculated from the corresponding fluxes (10-16), and expression (17) into the equation of motion (9), we obtain

$$\frac{d\overleftarrow{V}}{dt} = -\frac{1}{\rho}Div\overleftrightarrow{P} + \overleftarrow{F}_{ext} + \frac{2\overleftarrow{V}}{n\mu H}\left[\overleftarrow{E}\left(\overleftarrow{j}_{gr} - \overleftarrow{j}_{m} + \frac{\nu_{\perp}^{2}}{\nu_{II}^{2}}\overleftarrow{j}_{c} + \frac{\nu_{\perp}^{2}}{\nu_{II}^{2}}\overleftarrow{j}_{II}\right) + \right.$$
$$\left. + \frac{\nu_{\perp}^{2}}{e\nu_{II}^{2}}\overleftarrow{F}_{m}\left(\overleftarrow{j}_{c} + +\overleftarrow{j}_{II}\right) + \frac{1}{e}\left(\overleftarrow{F}_{c}\overleftarrow{j}_{m}\right)\right] + \tilde{\varepsilon}\frac{2V}{n\mu H}\frac{2H^{2}}{H^{2}+4\pi nmc^{2}}\left(\overleftarrow{F}_{m}\overleftarrow{j}_{p}\right), \tilde{\varepsilon} = (m_{e}/m_{i}) < <1. \tag{18}$$

Since the derived equation uses macroscopic quantities $n, \overleftrightarrow{P}, \overleftarrow{E}, \overleftarrow{H}, \overleftarrow{V}$ as the main parameters, there is no need for additional assumptions about the form of the distribution function associated with the termination of the chain of moments and the transition to hydrodynamic equations from the Vlasov kinetic equation. However, the most important thing in Eq. (18) is that it takes into account small dissipative and fluctuation processes arising due to the interaction of drift currents with inhomogeneous electric and magnetic fields. The reason for the smallness of the fluctuations taken into account in (18) is the condition of applicability of the leading center approximation and is a consequence of the perturbation theory, which is valid up to the constancy of the first adiabatic invariant ($\mu = const$)and therefore allows the parameters to vary within this accuracy. At the same time, it is known that fluctuations in plasma are responsible for the appearance of local currents, which are determined by space-time inhomogeneities in the distribution of the field and plasma. In turn, the interaction of these currents with forces, also associated with inhomogeneities in the spatial distribution of magnetic and electric fields, determines the further development of the resulting fluctuations, as well as the nature of the possible instability.

Eq. (18) under the assumption of quasineutrality ($n_e = n_i$) and infinite conductivity along the field line is greatly simplified ($\overleftarrow{E}_{II} = 0$). In addition, if we consider a closed axially symmetric system, then the inhomogeneity in the plasma distribution along the drift trajectory may be absent and the current intensity j_{II} proportional to this inhomogeneity tends to zero. Finally, instead of (18), we obtain a simplified, but not changing the physical essence, equation

$$\frac{d\overleftarrow{V}}{dt} = -\frac{1}{\rho}Div\overleftrightarrow{P} + \overleftarrow{F}_{ext} + \frac{2\overleftarrow{V}}{n\mu H}\left[\frac{\nu_{\perp}^{2}}{e\nu_{II}^{2}}\left(\overleftarrow{F}_{m}\cdot\overleftarrow{j}_{c}\right) + \frac{1}{e}\left(\overleftarrow{F}_{c}\cdot\overleftarrow{j}_{m}\right)\right] = \frac{1}{\rho}div\overleftrightarrow{P} + \overleftarrow{F}_{ext} + \overleftarrow{f}_{dis}\left(\overleftarrow{F}, \overleftarrow{j}\right) \tag{19}$$

Eqs. (18) and (19) differ from generally used equations of motion by the third term in the right part, which describes dissipative interaction of drift currents with $e\overleftarrow{E}, \overleftarrow{F}_m, \overleftarrow{F}_c$, forces. This additional part evidently take into account magnetization of physically infinitesimal element of a continuum, since besides the dependence on drift current $\overleftarrow{j}_{gr}, \overleftarrow{j}_{c}, \overleftarrow{j}_{II}$ and $\overleftarrow{j}_{m}$, it is proportional to $1/\mu$. We should note, that in the case with axial-symmetrical plasma system, currents $\overleftarrow{j}_{m}$ and $\overleftarrow{j}_{c}$ constantly flow in it. Nevertheless, they do not break freezing-in, as $\overleftarrow{j}_{m}$ and $\overleftarrow{j}_{c}$ are directed along the azimuth and $\overleftarrow{F}_m \perp \overleftarrow{j}_{c}$, $\overleftarrow{F}_c \perp \overleftarrow{j}_{m}$. At the same time the appearance of fluctuations may cause azimuthal inhomogeneity and, consequently, coincidence of $\overleftarrow{F}_m$ and $\overleftarrow{j}_{c}$, $\overleftarrow{F}_c$ and $\overleftarrow{j}_{m}$ components. Moreover, $\overleftarrow{F}_{ext} \sim \left[\overleftarrow{j}, \overleftarrow{H}\right]$ force in the Eqs. (18) and (19) is expressed through drift current explicit values, not through $rot\overleftarrow{H}$, which is within the framework of general conception of this chapter: consideration of current corpuscular structure.

4. Dissipative system of equations in the approximation of two adiabatic invariants of Chu, Goldberger, Low in the drift approximation.

In order to obtain a complete system of hydrodynamic equations in the drift approximation, it is necessary to add Maxwell's equations to the equation of motion (19), and to close the system, add two equations of state for the parallel ρ_{II} and perpendicular $\rho_{\perp}$ components of the pressure tensor, as is done in the approximation of two adiabatic invariants of the ChGL [14]. If one equation for is a consequence of the applicability of the drift approximation and corresponds to the constancy of the first adiabatic invariant $(d\mu/dt) = 0$, then the second equation can be obtained on the basis of the energy conservation law in the drift approximation [17]

$$\frac{d\varepsilon}{dt} = e\left(\overleftarrow{E} \cdot \overleftarrow{U}_{dr}\right) + \mu\frac{\partial H}{\partial t}, \tag{20}$$

where $\varepsilon = \varepsilon_{II} + \varepsilon_{\perp} = \left(mv_{II}^2/2\right) + \left(mv_{\perp}^2/2\right)$ is particle mean energy, U_{dr} is drift velocity. From (20) we obtain

$$\frac{d\varepsilon_{II}}{dt} = e\left(\overleftarrow{E} \cdot \overleftarrow{U}_{dr}\right) + \mu\frac{\partial H}{\partial t} - \frac{d\varepsilon_{\perp}}{dt} = e\left(\overleftarrow{E} \cdot \overleftarrow{U}_{dr}\right) - \mu\left(\overleftarrow{U}_{dr} \cdot \overleftarrow{\nabla}\right)H. \tag{21}$$

Since

$$\frac{d(n\varepsilon_{II})}{dt} = \varepsilon_{II}\frac{dn}{dt} + n\frac{d\varepsilon_{II}}{dt}$$

and $p_{II} = 2n\varepsilon_{II}$, $p_{\perp} = n\varepsilon_{\perp}$ and $\dot{n} = -ndiv\overleftarrow{U}_{dr}$ are valid, than from (20,21) and the latest expression we obtain

$$\frac{dp_{II}}{dt} = 2ne\left(\overleftarrow{E} \cdot \overleftarrow{U}_{dr}\right) - \frac{2p_{\perp}}{H}\left(\overleftarrow{U}_{dr} \cdot \overleftarrow{\nabla}\right)H - p_{II}div\overleftarrow{U}_{dr}$$

or

$$\frac{dp_{II}}{dt} + p_{II}div\overleftarrow{U}_{dr} = 2ne\left(\overleftarrow{E} \cdot \overleftarrow{U}_{dr}\right) - \frac{2p_{\perp}}{H}\left(\overleftarrow{U}_{dr} \cdot \overleftarrow{\nabla}\right)H. \tag{22}$$

Relation (22) is a substantial balance equation in the drift approximation for the pressure tensor component ρ_{II} with a nonzero right-hand side (the presence of a source). We multiply the left-hand side of (22) by H^2/ρ^3 and, taking into account that $\rho div\overleftarrow{U}_{dr} = -(d\rho/dt)$, we obtain after transformations

$$\frac{H^2}{\rho^3}\left(\frac{dp_{II}}{dt} - \frac{p_{II}}{\rho}\frac{d\rho}{dt}\right) = \frac{H^2}{\rho^3}\frac{dp_{II}}{dt} + \frac{p_{II}H^2}{\rho^2}\frac{d}{dt}\left(\frac{1}{\rho}\right) = \frac{H^2}{\rho^3}\frac{dp_{II}}{dt} + p_{II}\frac{d}{dt}\left(\frac{H^2}{\rho^3}\right) = \frac{d}{dt}\left(\frac{p_{II}H^2}{\rho^3}\right).$$

Now, after multiplying the right part (22) by $\left(H^2/\rho^3\right)$, we equate this product to the latest equation. Finally, we obtain

$$\frac{d}{dt}\left(\frac{p_{II}H^2}{\rho^3}\right) = \frac{H^2}{\rho^3}\left[2ne\left(\overleftarrow{E}\cdot\overleftarrow{U}_{dr}\right) - \frac{2p_\perp}{H}\left(\overleftarrow{U}_{dr}\cdot\overleftarrow{\nabla}\right)H\right]. \tag{23}$$

The condition

$$\frac{d}{dt}\left(\frac{p_\perp}{\rho H}\right) = 0, \tag{24}$$

equivalent to the condition of the first adiabatic invariant conservation, (since $\overline{v}_\perp^2 \approx p_\perp/\rho$, where $\overline{v}_\perp$ is a perpendicular component of particle mean velocity), together with (23) are two condition equations for the parallel p_{II} and $p_\perp$ perpendicular components of pressure tensor, which close the dissipative system of equations in drift approximation.

Now, the first part of (23) is under analysis. Since we consider plasma systems in axial-symmetrical magnetic fields with potential electric field equal to zero, than in the stationary case $E = 0$, $\left(\overleftarrow{U}_{dr}\cdot\overleftarrow{\nabla}\right)H \sim U_\varphi(\partial H/\partial\varphi) = 0$ and the right part (23) identically vanish. In variable fields, in our case $\left(\overleftarrow{E}\overleftarrow{U}_{dr}\right) = E_\varphi U_\varphi$, since for electric field $\overleftarrow{E}_\varphi = -\frac{1}{c}\frac{\partial\overleftarrow{A}_\varphi}{\partial t}$ and

$$\left(\overleftarrow{U}_{dr}\cdot\overleftarrow{\nabla}\right)H \approx \overleftarrow{U}_R\frac{\partial H}{\partial R} = \frac{U_R H}{R_{cr}} = \frac{U_R H}{R}k,$$

where k is a coefficient of proportionality between field line curvature radius R_{cr} and guiding center radius-vector R [18, 19], $U_R = c(E/H)$ is electric drift velocity. In the result, for the right part of (23) we obtain

$$2\frac{H^3}{\rho^3}n\cdot U_\varphi\cdot U_R\left(\frac{e}{c}n - \frac{\mu\cdot k}{R\cdot U_\varphi}\right).$$

According to the results of the papers [18, 19], we have

$$\frac{RU_\varphi}{k} = \frac{c}{e}\mu + \frac{v_{II}^2}{c\omega_L} = \frac{c}{enH}\left(p_\perp + p_{II}\right)$$

and, finally, for (23)) we may write

$$\frac{d}{dt}\left(\frac{p_{II}H^2}{\rho^3}\right) = \left(\frac{p_{II}H^2}{\rho^3}\right)\cdot\frac{2nU_\varphi}{p_\perp + p_{II}}\cdot eE_\varphi, \tag{25}$$

The total equation system of two adiabatic invariant approximations, considering $\overleftarrow{f}_{dis}$ in the approximation of ideal conductivity $E_\varphi \sim \left[\overleftarrow{V},\overleftarrow{H}\right]$, is written as follows:

$$\begin{gathered}\frac{d\overleftarrow{V}}{dt} = -\frac{1}{nm}Div\overleftrightarrow{P} + \overleftarrow{F}_{ext} + \overleftarrow{f}_{dis}\left(\overleftarrow{F},\overleftarrow{j}\right), \frac{\partial n}{\partial t} = -\overleftarrow{\nabla}\left(n\overleftarrow{V}\right), \frac{d}{dt}\left(\frac{p_\perp}{\rho H}\right) = 0,\\ \frac{d}{dt}\left(\frac{p_{II}H^2}{\rho^3}\right) = \left(\frac{p_{II}H^2}{\rho^3}\right)\cdot\frac{2nU_\varphi}{p_\perp + p_{II}}\cdot eE_\varphi, \frac{\partial\overleftarrow{H}}{\partial t} = rot\left[\overleftarrow{V},\overleftarrow{H}\right], \overleftarrow{E}_\varphi = \frac{1}{c}\left[\overleftarrow{V},\overleftarrow{H}\right],\end{gathered} \tag{26}$$

where - $(P)_{kn} = p_{II}\overleftarrow{e}_k\overleftarrow{e}_n + p_\perp\left(\delta_{kn} - \overleftarrow{e}_k\overleftarrow{e}_n\right), \overleftarrow{F}_{ext} = \frac{1}{H}\left[-p_\perp\overleftarrow{\nabla}_{II}H + (p_{II} - p_\perp)\overleftarrow{\nabla}_\perp H\right],$
$\overleftarrow{f}_{dis} = \frac{2cV}{enH^3}\left[2\frac{p_\perp}{R^2}\left(\overleftarrow{\nabla}H\left[\overleftarrow{R}\overleftarrow{H}\right]\right) - p_{II}\left(rot\overleftarrow{H}\cdot\overleftarrow{\nabla}_\perp H\right)\right].$

Let us multiply the first equation in the system (26) scalarly by V

$$\left(\overleftarrow{V}\cdot\frac{d\overleftarrow{V}}{dt}\right) = -\frac{1}{nm}\left(\overleftarrow{V}\cdot Div\overleftrightarrow{P} + \left(\overleftarrow{V}\cdot\overleftarrow{F}_{ext}\right) + \left(\overleftarrow{V}\cdot\overleftarrow{f}_{dis}\left(\overleftarrow{F},\overleftarrow{j}\right)\right)\right)$$

Since we are interested in the influence of the dissipative term on the character of motion of a plasma element with macroscopic velocity $\overleftarrow{V}$, let us assume for simplicity that the scalar product of the first two terms is early to zero, then, given the explicit form $\overleftarrow{f}_{dis}$, we obtain

$$\frac{1}{2}\frac{dV^2}{dt} = \frac{2cV^2}{enH^3}\left[2\frac{P_\perp}{R^2}\left(\overleftarrow{\nabla}H\left[\overleftarrow{R},\overleftarrow{H}\right]\right) - p_{II}\left(rot\overleftarrow{H}\cdot\overleftarrow{\nabla}_\perp H\right)\right].$$

The last expression shows that in the case of fluctuations, azimuthal inhomogeneity may appear and, as a consequence, to the coincidence of the direction of the components $\overleftarrow{F}_m$ and $\overleftarrow{j}_c$, and $\overleftarrow{j}_m$ (see (19) and explanations to it), then, depending on the sign of the term in square brackets, the energy of the plasma element will increase or change.

5. Conclusions

The right parts of the functions $\overleftarrow{F}_{ext}$ and $\overleftarrow{f}_{dis}$ are expressed through drift current explicit values separating the components of pressure tensor $p_\perp$ and p_{II}. In the system (26) the unknown values are $p_\perp, p_{II}, \overleftarrow{H}, \overleftarrow{E}_\varphi, n$ and $\overleftarrow{V}$.

Obtaining theoretical models describing the motion of continuous systems is an important branch of continuum mechanics. The construction of these models is based both on the use of experimental data and on the application of the well-known principles of mechanics, thermodynamics, physics, and they are based on the search for additional relationships between the parameters describing the state of the considered continuous medium. It is known that the basic equations of mechanics, electrodynamics, hydrodynamics, and so on are derived on the basis of the variational Lagrange equation. The corresponding analysis shows that with the help of variational principles it is possible to construct any physical models describing both reversible and non-reversible processes. Therefore, the application of the principles of Prigogine and Onsager, combined by Gyarmati, to obtain the equation of motion of a magnetized plasma at the hydrodynamic level of description seems to be quite promising. And here the following should be noted.

In the hydrodynamic approximation, fluctuations are not taken into account, since in continuum mechanics it is assumed to be continuous. The Navier-Stokes equation, in contrast to the Euler equation, already takes into account dissipative phenomena, but does not contain fluctuation interactions (without additional assumptions about the form of the stress tensor that takes into account the molecular structure), which describe Brownian motion. Relying on the concept of continuity, as already noted, it is impossible in the mechanics of continuous media to take into account the fluctuations of the hydrodynamic functions formed due to the molecular structure of the medium. At this level of description, taking into account the atomic-molecular structure leads to the Langevin equation, in which the parameters of the medium are described by random sources. These sources are responsible for fluctuations $\rho, \overleftarrow{V}, T$ and being unavoidable properties of the medium, cannot be excluded. Therefore, postulating Langevin sources in hydrodynamics brings the corresponding equations as close as possible to describing the behavior of a real medium.

In turn, the possibility of taking into account the structure of a physically infinitesimal plasma element in this work was achieved, on the one hand, by using the variational methods of Prigogine and Onsager, combined by Gyarmati, and making it possible to obtain a completely self-consistent equation with the accuracy of the chosen drift approximation. On the other hand, this approximation, being single-particle, initially takes into account the discreteness of the considered ionized medium ("atomic-molecular" structure). In addition, it admits small perturbations within the limits of its accuracy, that is, within the limits of the constancy of the first adiabatic invariant μ.

1. The application of the methods of nonequilibrium thermodynamics based on the combined principles of Prigogine and Onsager for the description in the linear approximation of transport processes in a collisionless plasma and taking into account the structure of a physically infinitesimal element of the medium ($\ell_f \sim \lambda_D$) makes it possible to obtain the equation of motion of an electron-ion plasma in the drift approximation. This equation takes into account fluctuation-dissipative processes, which are determined by the interaction of local drift currents and forces. The expediency of an approach in which the influence of local currents is taken into account in describing the behavior of a nonequilibrium plasma was noted in [20]. The resulting fluctuations lead to the formation of spatial inhomogeneities in the distribution of the field and plasma and to the coincidence of the components of the current and forces and "turn on" the dissipative source, which determines the further development of possible instabilities.

2. The transition from an arbitrary to an axially symmetric magnetic system greatly simplifies the equation of motion, but retains the basis associated with taking into account the structure of a physically infinitesimal element. This is fundamental in comparison with the usual nondissipative Euler equation, which is used in a one-fluid hydrodynamic system of equations and in the system of equations of two adiabatic invariants of Chu, Goldberger, and Low.

3. The possibility of kinetic foundation of the postulated in the paper random source in pressure nonequilibrium part appears, when small scale initial correlations, which are superimposed for derivation of Landau and Vlasov equations, partially decreasing.

4. The obtained equation of motion can be used when taking into account the scattering of charged particles by electromagnetic fluctuations. This determines an additional mechanism for the regularization of particle motion in a magnetized plasma, which automatically implies a revision of the scale associated with the path length determined by the Coulomb collision, since it may turn out to be much

larger than the distance between collisions on fluctuations. For example, in the problem of plasma flow around the solar wind of the Earth's magnetosphere, the characteristic size of the latter is much less than the mean free path corresponding to Coulomb collisions. This, proceeding from rigorous considerations, indicates the inadmissibility of using the hydrodynamic approximation to describe the processes in this problem. However, the experimental data are in good agreement with the results that follow for this problem from the solution of hydrodynamic equations, which indicates the presence of an effective particle scattering mechanism, which leads to a significant decrease in the mean free path in comparison with Coulomb collisions [12].

Abbreviations

T_L	the period of Larmor's rotation
ℓ_f	scale, characterizing a physically small "point" of a solid medium
g	plasma parameter
n	particle concentration
λ_D	Debye radius
H	magnetic field strength
$\overleftarrow{V}$	macroscopic average plasma velocity
P	pressure tensor
$\overleftarrow{j}$	current density
ρ_{L_i}	Larmor's ion radius
ρ_{L_e}	Larmor's electron radius
L	lagrangian
ω	frequency
ω_{L_i}	Larmor frequency of the ion
ω_{0_i}	plasma ion frequency
τ_f	characteristic time
N_f	number of particles in volumel_f^3
V_T	thermal velocity of the particle
σ	entropy production function
J_i	fluxes corresponding to the observed transfer processes
X_i	thermodynamic forces
Ψ	scattering potentials
Φ	scattering potentials
$L_{i,k}$	Onsager's reciprocity coefficient
$R_{i,k}$	inverse Onsager reciprocity coefficient
U	volume
$\overleftrightarrow{P}_d$	nonequilibrium part of the pressure tensor
$\overleftrightarrow{P}^{i,e}$	equilibrium part of the pressure tensor of the ionic and electron components
$\overleftrightarrow{P}_d^{i,e}$	nonequilibrium part of the pressure tensor of the ion and electron components
$\overleftrightarrow{P}_\Sigma^{i,e}$	total pressure for the electron and ion component

$m_{i,e}$	mass of an ion or electron
$\overset{\leftrightarrow}{I}$	unit tensor
$\overset{\leftarrow}{V}_{i,e}$	macroscopic velocity of the ion and electron components
$p^{i}_{II,\perp}$	parallel and perpendicular pressure components
$\overset{\leftarrow}{e}_k$	unit vectors
$\overset{\leftarrow}{e}_n$	unit vectors
$\overset{\leftarrow}{e}_1 = \overset{\leftarrow}{H}/H$	a unit vector pointing along the field
$\rho_{i,e}$	density of the ion and electron components of the plasma
$\overset{\leftarrow}{F}{}^{i,e}_{ext}$	external force acting on electrons and ions
$T_{i,e}$	temperature of the ion and electron plasma components
$\overset{\leftarrow}{V}_{i,e}$	average velocity of the ionic and electronic components
$\nu_{II,\perp}$	perpendicular and parallel components of the particle velocity
$\overset{\leftarrow}{j}_m$	magnetizing current
$\overset{\leftarrow}{j}_{gr}$	gradient drift current
$\overset{\leftarrow}{R}$	radius-vector of the particle
$\overset{\leftarrow}{F}_m$	magnetic force
$\overset{\leftarrow}{F}_c$	centripetal force
$\overset{\leftarrow}{j}_c$	centripetal current
$\overset{\leftarrow}{j}_{II}$	parallel current
μ	magnetic moment magnetic moment
$\overset{\leftarrow}{j}_p$	polarizing current
ε	particle energy
U_{dr}	drift velocity
E_{φ}	azimuthal component of the electric field
U_{φ}	azimuthal drift velocity
k	coefficient of proportionality between the radius of curvature of the force line and the radius vector
R_{cr}	radius of curvature of the force line
U_R	radial drift velocity
$\overset{\leftarrow}{f}_{dis}$	dissipative force
$\tilde{\varepsilon}$	order of smallness
$\overset{\leftarrow}{A}_{\varphi}$	azimuthal component of the vector potential

Appendix

Let us calculate the divergences from each of the fluxes $J_2 - J_7$, whose explicit form is represented by expressions (10a)–(10g) $(divJ_1 = 0)$. In the preconversion process we will take into account the invariance of n/H and the first adiabatic invariant $/mu = \left(m\nu_{\perp}^{2}/2H\right)$, as well as the corresponding preconversion of the divergence from the vector product $div[A,B] = (B \cdot rotA)(A \cdot rotB)$.

For the flow divergence J_2 we obtain

$$div\overleftarrow{J}_2 = div\left(nc\frac{\left[\overleftarrow{E},\overleftarrow{H}\right]}{H^2}\right) = \frac{nc}{H^2}div\left[\overleftarrow{E},\overleftarrow{H}\right] + \left[\overleftarrow{E},\overleftarrow{H}\right]div\left(\frac{nc}{H^2}\right) = -\frac{nc}{H^2}\left(\overleftarrow{E}rot\overleftarrow{H}\right) -$$

$$-\frac{nc}{H}\left(\left[\overleftarrow{E},\overleftarrow{H}\right]\overleftarrow{\nabla}\frac{1}{H}\right) = \frac{2}{mv_{\perp}^2}\left(\overleftarrow{E}\overleftarrow{j}_m\right) - \frac{nc}{H^3}\overleftarrow{E}\left[\overleftarrow{H},\overleftarrow{\nabla}H\right] = \frac{2}{mv_{\perp}^2}\left(\overleftarrow{E}\overleftarrow{j}_m\right) - \frac{2}{mv_{\perp}^2}\left(\overleftarrow{E}\overleftarrow{j}_{gr}\right), \tag{27}$$

where

$$\overleftarrow{j}_m = -\frac{nc}{H}\mu rot\overleftarrow{H}, j_{gr} = \frac{nc}{H^2}\mu\left[\overleftarrow{H},\overleftarrow{\nabla}H\right].$$

Let's calculate the divergence from the J_3 flow:

$$div\overleftarrow{J}_3 = div\left(n\frac{mcv_{\perp}^2}{2eH^3}\left[\overleftarrow{H},\overleftarrow{\nabla}H\right]\right) = n\frac{mcv_{\perp}^2}{2eH^3}div\left[\overleftarrow{H},\overleftarrow{\nabla}H\right] + \left[\overleftarrow{H},\overleftarrow{\nabla}H\right]\overleftarrow{\nabla}n\frac{mcv_{\perp}^2}{2eH^3} =$$

$$= -\frac{p_{\perp}}{eH^3}\overleftarrow{\nabla}Hrot\overleftarrow{H} - \left[\overleftarrow{H},\overleftarrow{\nabla}H\right]n\frac{mcv_{\perp}^2}{2eH^2}\overleftarrow{\nabla}\frac{1}{H} = \tag{28}$$

$$= \frac{\left(\overleftarrow{j}_m\overleftarrow{\nabla}\right)H}{eH} + n\frac{mcv_{\perp}^2}{2eH^4}\left[\overleftarrow{H},\overleftarrow{\nabla}H\right] = \frac{2}{mev^2}\left(\overleftarrow{F}_m\overleftarrow{j}_m\right),$$

The final result in formulas (27) and (28) correspond to formulas (11) and (13). where $\overleftarrow{F}_m = -\mu\overleftarrow{\nabla}H$.

Similarly, transform the divergences from the fluxes $J_4 - J_7$, we obtain

$$div\overleftarrow{J}_4 = -\frac{2}{mev_{\perp}^2}\left(\overleftarrow{j}_m\overleftarrow{\nabla}\frac{p_{\perp}}{n}\right) + \frac{2}{mev_{\perp}^2}\left(\overleftarrow{j}_{gr}\overleftarrow{\nabla}\frac{p_{\perp}}{n}\right). \tag{29}$$

$$div\overleftarrow{J}_5 = -\frac{2}{mev_{\perp}^2}\left(\overleftarrow{F}_c\overleftarrow{j}_m\right) + \frac{2}{mev_{II}^2}\left(\overleftarrow{j}_c\overleftarrow{\nabla}\frac{p_{II}}{n}\right), \tag{30}$$

$$div\overleftarrow{J}_6 = -\frac{2}{mev_{\perp}^2}\left(\overleftarrow{j}_m\overleftarrow{\nabla}\frac{p_{II}}{n}\right) + \frac{2}{mev_{\perp}^2}\left(\overleftarrow{j}_{gr}\overleftarrow{\nabla}\frac{p_{II}}{n}\right), \tag{31}$$

$$div\overleftarrow{J}_7 - -\frac{n}{H}\left(\overleftarrow{H}\overleftarrow{\nabla}v'_{II}\right) - \left(\overleftarrow{e_1}\overleftarrow{\nabla}v'_{II}\right) \tag{32}$$

In (29)-(32) it is necessary to transform the gradient terms from and. To do this, we use the invariance of $(p_{\perp}/n)$, (p_{II}/n) and v_{II}. Since

$$\overleftarrow{\nabla}\frac{p\perp}{n} = \overleftarrow{\nabla}\frac{mv_{\perp}^2}{2} = \overleftarrow{\nabla}\frac{mv_{\perp}^2}{2H}H = \mu\overleftarrow{\nabla}H = -\overleftarrow{F}_m,$$

Then

$$\left(\overleftarrow{j}_m\overleftarrow{\nabla}\frac{p_{\perp}}{n}\right) = -\left(\overleftarrow{j}_m\overleftarrow{F}_m\right) \tag{33}$$

and

$$\left(\overleftarrow{j}_{gr}\overleftarrow{\nabla}\frac{p_{\perp}}{n}\right) = -\left(\overleftarrow{j}_{gr}\overleftarrow{F}_m\right). \tag{34}$$

In a stationary magnetic field it is true with the accuracy of the drift approximation [21]

$$\frac{d\nu_{II}}{dt} = \frac{e}{m}\left(\overleftarrow{E}\overleftarrow{e}_1\right) + \frac{\nu_{\perp}^2}{2} div\overleftarrow{e}_1. \tag{35}$$

Assume that the first term in (35) is zero (the electric field is perpendicular to the magnetic field). Convert the second term in (35)

$$\frac{\nu_{\perp}^2}{2} div\overleftarrow{e}_1 = \frac{\nu_{\perp}^2}{2} div\frac{\overleftarrow{H}}{H} = \frac{\nu_{\perp}^2}{2} div\overleftarrow{H} - \frac{\nu_{\perp}^2}{2H^2}\left(\overleftarrow{H}\overleftarrow{\nabla}H\right) = -\frac{\mu}{m}\left(\overleftarrow{e}_1\overleftarrow{\nabla}H\right) = -\frac{1}{m}\left(\overleftarrow{e}_1\overleftarrow{F}_m\right).$$

Given this transformation (35) will take the form

$$\frac{d\nu_{II}}{dt} = \frac{1}{m}\left(\overleftarrow{e}_1\overleftarrow{F}_m\right). \tag{36}$$

In addition, for the constant magnetic field in the drift approximation it is true

$$\frac{d\nu_{II}}{dt} = \frac{\partial\nu_{II}}{\partial t} + \left(\overleftarrow{U}\overleftarrow{\nabla}\right)\overleftarrow{\nabla}\nu_{II} = \left(\nu_{II}\overleftarrow{e}_1 + \overleftarrow{U}_{dr.}\right)\overleftarrow{\nabla}\nu_{II} \approx \nu_{II}\left(\overleftarrow{e}_1\overleftarrow{\nabla}\nu_{II}\right). \tag{37}$$

By equating (36) and (37), we obtain

$$\overleftarrow{e}_1\nu_{II}\overleftarrow{\nabla}\nu_{II} = \frac{1}{m}\left(\overleftarrow{e}_1\overleftarrow{F}_m\right). \tag{38}$$

Let's write the gradient from

$$\overleftarrow{\nabla}\left(\frac{p_{II}}{n}\right) = \overleftarrow{\nabla}m\nu_{II}^2 = 2m\nu_{II}\overleftarrow{\nabla}\nu_{II},$$

whence, taking into account, we have

$$\overleftarrow{j}_{gr.}\overleftarrow{\nabla}\frac{p_{II}}{n} = 2e\left(\overleftarrow{E}\overleftarrow{j}_{gr.}\right) + 2\left(\overleftarrow{F}_m\overleftarrow{j}_{gr.}\right), \tag{39}$$

$$\overleftarrow{j}_{m.}\overleftarrow{\nabla}\frac{p_{II}}{n} = 2e\left(\overleftarrow{E}\overleftarrow{j}_{m.}\right) + 2\left(\overleftarrow{F}_m\overleftarrow{j}_{m.}\right), \tag{40}$$

$$\overleftarrow{j}_{c.}\overleftarrow{\nabla}\frac{p_{II}}{n} = 2e\left(\overleftarrow{E}\overleftarrow{j}_{c.}\right) + 2\left(\overleftarrow{F}_m\overleftarrow{j}_{c.}\right), \tag{41}$$

$$\left(\overleftarrow{e}_1\overleftarrow{\nabla}\nu_{II}\right) = \frac{e}{m\nu_{II}}\left(\overleftarrow{E}\overleftarrow{e}_1\right) + \frac{\left(\overleftarrow{e}_1\overleftarrow{F}_m\right)}{m\nu_{II}}, \tag{42}$$

By substituting the values of (33), (34), (39)-(42), into (29)-(32), we obtain the expressions (13), (14), (15) and (16) presented in Section 3, respectively.

If a time-varying electric field acts in the plasma, the crossed $\overleftarrow{E}, \overleftarrow{H}$ fields produce an acceleration of electric drift

$$\frac{d\overleftarrow{V}_E}{dt} = c\frac{\left[\overleftarrow{E}, \overleftarrow{H}\right]}{H^2} \tag{43}$$

creating an inertial force $\overleftarrow{F}_{iner.} = -m\,\dot{\overleftarrow{\nu}}_E$.

In the drift approximation, the electric field $\overleftarrow{E}$ and its rate of change are limited by $(cE/H) < < V$ and $(\partial E/\partial t < < E/T_L)$ (T_L is the period of Larmor's rotation).

The force (43) causes drift with speed

$$\overleftarrow{\nu}_P = \frac{mc^2}{H^2}\dot{\overleftarrow{E}} \tag{44}$$

and leads to the occurrence of electric polarization current

$$\overleftarrow{j}_P = ne\overleftarrow{\nu}_P = \frac{nmc^2}{H^2}\dot{\overleftarrow{E}} \tag{45}$$

In (44) and (45) we took into account the equality to zero of the scalar product $\left(\overleftarrow{E} \cdot \overleftarrow{e}_1\right)$. According to (45) we have

$$\frac{\partial\overleftarrow{E}}{\partial t} = \frac{H^2}{nmc^2}\overleftarrow{j}_P. \tag{46}$$

Since

$$(n_e - n_i) = \frac{1}{4\pi e}div\,\dot{\overleftarrow{E}},$$

then

$$\frac{\partial}{\partial t}(n_e - n_i) = \frac{1}{4\pi e}div\frac{\partial\overleftarrow{E}}{\partial t}. \tag{47}$$

Substituting the values of from (46) into (47), we obtain

$$\frac{\partial(n_e - n_i)}{\partial t} = \frac{1}{4\pi e}div\left[\frac{H^2}{nmc^2}\overleftarrow{j}_P\right].$$

Calculating the divergence from the expression in square brackets of the last expression gives

$$\frac{\partial}{\partial}(n_e - n_i) = \frac{H^2}{4\pi emc^2}div\overleftarrow{j}_P - \frac{H^2}{\pi nemc^2}\frac{1}{m\nu_\perp^2}\left(\overleftarrow{j}_P\overleftarrow{F}_m\right). \tag{48}$$

In the derivation of (48) the spatial derivatives of $\overleftarrow{E}$ were neglected with the accuracy of the drift approximation. Let us now calculate the value of $div\overleftarrow{j}_P$. To do this, we substitute in Maxwell's equation

$$rot\overleftarrow{H} = \frac{1}{c}\frac{\partial \overleftarrow{E}}{\partial t} + \frac{4\pi}{c}\overleftarrow{j}_P$$

value of the current $\overleftarrow{j}_P$ from (45) and after simple transformations we obtain

$$rot\overleftarrow{H} = \frac{æ}{c}\overleftarrow{E},$$

where $æ = 1 + \frac{4\pi nmc^2}{H^2}$ [16]. From which we get

$$\frac{\partial \overleftarrow{E}}{\partial t} = \frac{c}{æ} rotH. \tag{49}$$

Substituting the value of the derivative according to (49) and (45), we obtain

$$\overleftarrow{j}_P = \frac{nmc^3}{(H^2 + 4\pi nmc^2)} rot\overleftarrow{H}. \tag{50}$$

Let's calculate the divergence from the right and left parts of (50), we get

$$div\overleftarrow{j}_P = \frac{4}{mv_\perp^2}\left(\overleftarrow{j}_P \cdot \overleftarrow{F}_m\right). \tag{51}$$

After substituting in (48) the value of, according to (51), we finally obtain expression (17) for the derivative of the concentration difference $(n_e - n_i)$ given in Section 3,

$$\frac{\partial}{\partial t}(n_e - n_i) = -\frac{4H^2}{(H^2 + 4\pi nmc^2)} \frac{\left(\overleftarrow{j}_P \overleftarrow{F}_m\right)}{emv^2}.$$

Author details

Vadim Bogdanov
Institute of Cosmophysical Research and Radio Wave Propagation, Far Easten Branch, Russian Academy of Sciences, Russia

*Address all correspondence to: vbogd@ikir.ru

References

[1] Alfven H. Space Plasma. M.: Mir; 1983. p. 216

[2] Klimontovich YL. Statistical Physics. New York: Harwood Acfdemic Pablishers; 1986

[3] Volkov TF. Gidrodinamicheskoe opisanie sil'no razregennoi plasmy. Voprosy fizisici plazmy. Pod red. M.A. Leontovicha. M.: Atomizdat; 1964. pp. 3-19

[4] Petschekb HE. The physics of solar flares. In: Ness WH, editor. AAS-NASA Symposiom. Vol. **SP-50**. NASA; 1964. p. 425

[5] Onsaqer L. Reciprocal relations in irreversible processes. Physics Review. 1931;**37**:405

[6] Priqoqin I. Bull. Acad. Roy. Belq. Cl. Sci. 1945;**31**:600

[7] Gyarmati I. Neravnovesnaya termodinamica. Vol. **304**. M.: Mir; 1974

[8] Landau LD. Kineticheskoe uravnenie v sluchae culonovskogo vzaimodeystviya. ZhETF. 1937;7:154

[9] Vlasov AA. Vibratsonnye svoystva electronnogo gaza. ZhETF. 1938;**8**:291

[10] Klimontovich YL. Statistical Theory of Open Systems. Dordrecht: Kluwer Academic pablishes; 1995. p. 1

[11] Klimontovich YL. Statisticheskaya teoriya otkrytykh sistem. Vol. **2**. Đœ.: Yanus-K; 1999. p. 438

[12] Baranov VB, Krasnobaev KV. Gidrodinamicheskaya teoriya kosmicheckoy plazmy. M.: Nauka; 1977. p. 336

[13] Kadomtsev BB. Kollektivnye protsy v plazme. M.: Nauka; 1976. p. 340

[14] Chew G, Goldberger M, Low F. Proceedings of the Royal Society. 1956; **A236**:1204

[15] Nortrop T. The Adiabatic Motion of Charged Particles. New York-London-Sydney: Interscience Publishers, a division of John Wiley and Sons; 1963

[16] Frank-Kamenskii DA. Lectures po fizike plazmy. M.: Atomizdat; 1968. p. 288

[17] Morozov AI, Solovev LS. Dvizhenie zaryannykh chastits v elektromagnitnykh polyakh. Voprosy fiziki Plazmy. Vol. 2. Pod red. M.A. Leontovicha. M.: Atomizdat; 1963. pp. 177-261

[18] Bogdanov VV, Pletnev VD. K voprosy o tochnosti sokhraneniya tret'ego adiabaticheskogo invarianta dvizheniya zaryannykh chastits v aksial'no-simmetrichnykh polyakh. Cosmic Research. 1972,**10**(3).350-367

[19] Bogdanov VV, Pletnev VD. K voprosy o tochnosti sokhraneniya tret'ego adiabaticheskogo invarianta dvizheniya zaryannykh chastits v aksial'no-simmetrichnykh polyakh. Cosmic Research. 1972;**10**(4):528-531

[20] Fadeev VM, Kvardtskhava IF, Komarov NN. Yadernyy sintez. 1965;**5**: 202

[21] Bogolubov NN, Mitropolski UA. Asimtoticheskie metody v teorii nelineinyx colebanij. Nauka. 1974:504

Section 2

Solar Wind, Humans and Communication

Chapter 7

Introductory Chapter: The Sun and Its Phenomenal Material Flux

Yann-Henri Chemin

1. Introduction

The Sun is the closest (by far) of the stellar objects we can study. Yet, we remotely sense it only, by means of recording electromagnetic spectra emitted from its activity. Remote sensing studies permit to analyze the Sun in many different perspectives, according to the types of spectra the instrument focus is. Thermodynamics, hydrodynamics applied to plasma with magnetic fields are all needed to study the radiative, convective, and exo-atmospheric conditions of the Sun energy transport. Stellar objects of different characteristics have been observed for ages by astronomers, and many physical theories have been developed relating observations and life cycles (i.e., HR diagram and equations of stellar structure, respectively). Stellar oscillations [1], spherical harmonics, and resonance patterns analysis belong to geophysics and are now in common use to study and classify stars.

More recently, scientists have been able to evolve geophysics techniques to reach within layers of the Sun. Seismology is now having a branch dedicated to the Sun: helioseismology. Magneto-convective analysis relies on remote sensing and physics, modeling, and inverse modeling. The Sun's core is in the process of nuclear fusion, where quantum physics applies. Atomic physics permits us to understand the life cycle of stars and their creation of atoms of different Z number depending on the type of star evolution as well as hydrostatic equilibrium conditions and environments. The Coulomb barrier and the Gamow peak involve both atomic physics and quantum statistics. Random walks taking 1 million years for a photon to leave the core to the radiative and arrive at the convective zone of the Sun are found from combined fields of physics and stochastic/Markovian modeling science.

Technological sciences such as supercomputing, applied experimental physics, engineering are also necessary to tackle complex modeling, simulation, and experiments. Astronomical engineering is particularly of importance in this case, on Earth and in orbit (space science/engineering) to gather remote sensing data. Lastly, theoretical mathematicians, physicists, and scientists have all had parts in developing the sciences used in the Sun analysis today, as the most fundamental of the sciences reverberate in the different bits of understanding found in the group of sciences that evolved from simple timing of and logging of visible events, to more complex instrumentation/analysis as observation sciences became more technological. A final question that comes to mind when analyzing electromagnetic interaction of magnetism such as the Zeeman effect is where is the electric part in the research on the Sun's electromagnetism. Is there magnetism alone only in the Sun?

IntechOpen

2. Connection from the inner core to the surface

The central part of the Sun is composed of a core (about a fourth of its radius) where thermonuclear reactions generate energy. While its average density is about 10 times that of lead, its temperature is about 15 10^6K. The core yields to the radiative zone of the Sun, which is about one-third of the Sun's radius. Both the core and the radiative zone transfer energy by radiative forces of photons following a random walk and seem to act as an apparent solid body [2].

The photosphere is 300–500Km deep, it is the part of the Sun from which the light is being emitted, before the plasma becomes opaque. The effective temperature of the Sun comes from the photosphere, at about 5.8 10^3K, plasma convection is visible there under the form of granules of sizes measured in Mm (10^3m). As any convective cell, granules [3, 4] have central heat upwelling and peripherial cool downwelling (**Figure 1**). Their life span is short, 8–20 minutes. Within the granules area, additionally to pores and magnetic flux tubes, sunspots may rise within mid-latitudes and converge to subequatorial latitudes. A sunspot is a cooler region and is measured in tens of Mm. It is composed of a central darkest part, the umbra, and surrounded by a less dark area called the penumbra made of radial acicularity (**Figure 1**). Its existence is inherently due to the Sun's magnetism.

Further parts outside of the Sun are referred to as part of its "atmosphere." They are the chromosphere and the corona (in that order).

The chromosphere is 2000 km deep, it is the Sun's eclipse "red ring of fire." It is characterized by a steep drop in material density, and an initial temperature drop from 5.8 10^3K to 3.5 10^3K to eventually reach 35 10^3K. The chromosphere and its transition zone to the next zone are the subject of study of the Interface Region Imaging Spectrograph (http://www.nasa.gov/iris), especially the chromospheric jets associated with coronal heating (de Pontieu, 2011 @SETI Talks).

The corona is a very large volume above the chromosphere, vastly warmer too, made of ionized plasma of about 1 10^6K, with a majority of emission coming from Fe-XIV and Fe-X. It is the origin of the *solar winds*. Some areas with open magnetic fields (**Figure 2**) yield faster *solar winds* (about 0.7 10^6m/s). The Hinode mission

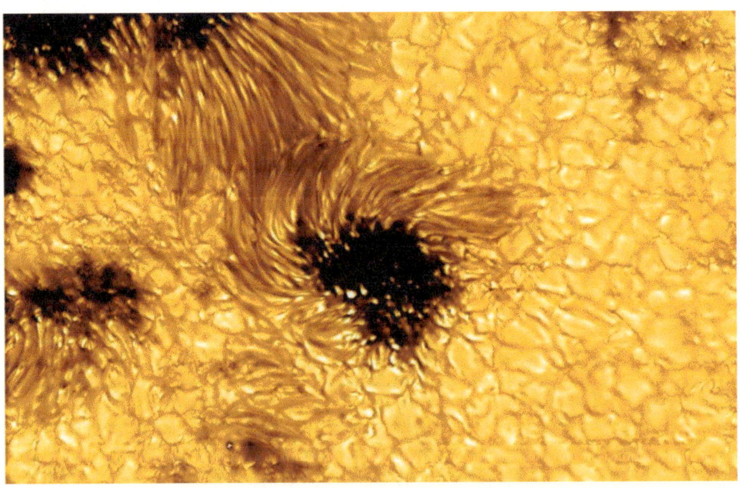

Figure 1.
Granule convection cells around sunspots. Credit: ESA & NASA/solar orbiter/EUI team.

Figure 2.
High-resolution image of the sun from solar orbiter, showing magnetically bound plasma. Credit: ESA & NASA/ solar orbiter/EUI team; data processing: E. Kraaikamp (ROB) [5].

recorded the most sensitive information about the magnetism of the Sun (http:// www.nasa.gov/hinode/).

The interplay between convection and magnetic fields drives all the heating in the solar atmosphere and the space weather. The magneto-convective energy heats the corona, drives the *solar wind*, causes flares and coronal mass ejections (de Pontieu, 2011 @SETI Talks).

Thompson et al. [6] used continuous observations from the Global Oscillation Network Group (GONG) and inverse modeling (harmonics frequency splitting, inverting rotation kernels) to confirm that differential rotation on the surface is carrying through most of the convection zone, until the tachocline at 0.713 of the Sun's atmosphere radius (R), a zone (**Figure 3**) of strong shear where disassociation happens with the deeper part of the Sun [2], and where variations of rotations have been linked with the presumed depth of the solar dynamo. Howe [2] further found that temporal variations in the tachocline region extend in the radiative zone as far as 0.63R (blue overlay on **Figure 3**), which suggests complex physics properties at the shear zone converting from (apparent) nearly solid state to convective.

Thompson et al. [6] also observed a shear layer just below surface at lower latitudes. In the equatorial regions, as depth increases, the rotation rate first increases (orange overlay on **Figure 3**) and then decreases. This particularity is reducing with latitude (away from equator). Sunspots are forced by the differential rotation to "stretch" from their roots in the convection zone and below [7] following the equatorial convergence until their "elasticity" is reaching limits. At this point, they "snap" to release as a filament.

A filament eruption (http://science.nasa.gov/missions/trace/) is a magnetic line disconnecting from the underlying magnetic field after a too large disturbance, the Alfvén waves survive the Corona transfer and are depositing energy in the *solar wind* [8].

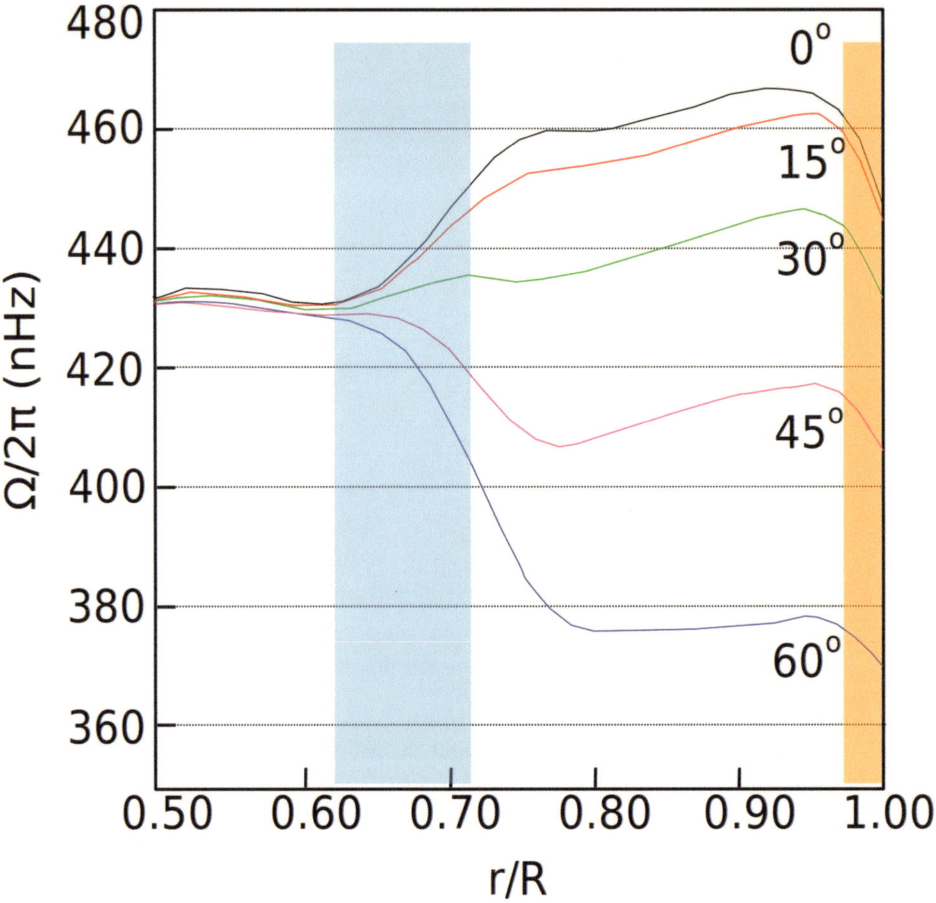

Figure 3.
Time-averaged rotation rates (Ω/2π) vs. partial sun radius (r/R) at different latitudes, redrawn from Howe [2].

3. The Sun convection zone and the processes of upwelling matter

Nandy et al. [7] studying the mass conservation of the Sun Convection Zone (SCZ) looked into modeling of flow transfer within meridional latitudes and inner tachocline structures. They found that the mass conservation must include toroidal fields in opposite direction at higher latitudes, to compensate the sunspot nursery at lower latitudes. What they also found is that the poloidal fields within the SCZ also drive mass transfer from equator to high latitudes. Their simulation also demonstrated that the mass transfer actually happens *through* the tachocline. It seems that the toroidal field is somehow passing through the tachocline at low latitudes (creating sunspots), but not its opposite direction equivalent at high altitudes (absence of sunspots). Is the tachocline generating a toroidal field constraint larger at high latitudes? It seems (**Figure 4**) that the combination of poloidal and toroidal field simulation adds explanations about the sunspots appearance in the lower latitudes, as the poloidal field/mass movement is outward at low latitude, and inward at high latitude [7]. In that condition, the threshold for sunspot generation used by Nandy et al. [7] is 10^5 G above the base of the SCZ. and tachocline (gray) redrawn from Nandy et al. [7].

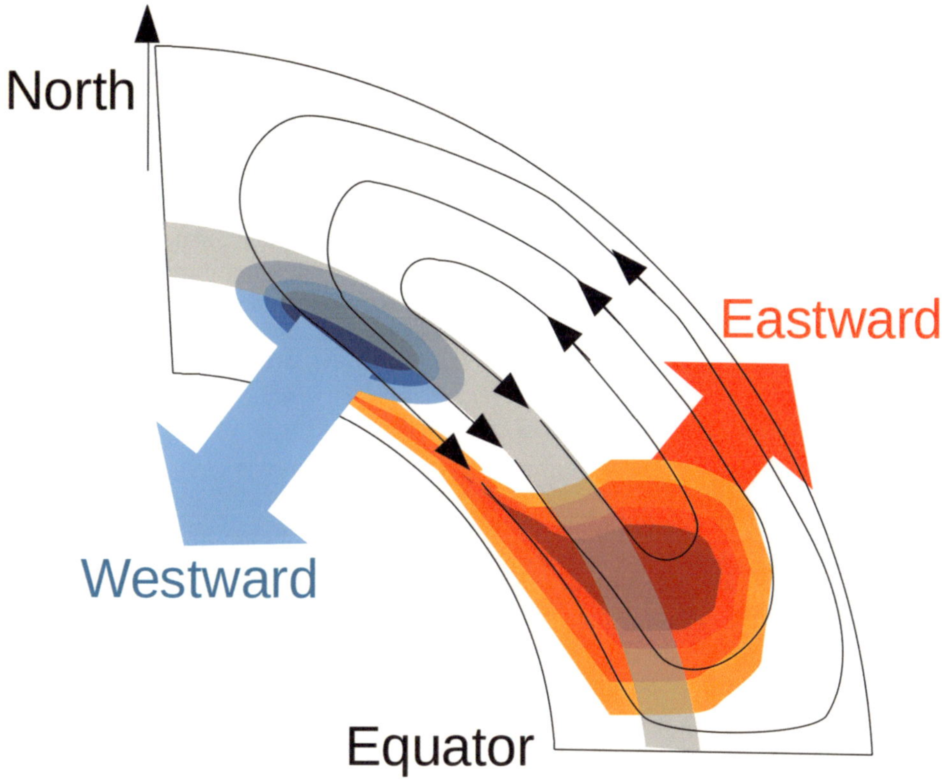

Figure 4.
Flow directions in toroidal fields (blue/red), poloidal fields (black lines), redrawn after [7].

Hathaway and Rightmire [9] studied the 1996–2009 period's magnetic maps made every 96 minutes from the MDI sensor (discontinued in 2011, follow-up by hmi.stanford.edu) on-board SOHO (soho.nascom.nasa.gov). MDI imaged the line-of-sight magnetic field by measuring the difference of polarization on both side of a Nickel absorption line in the Sun's atmosphere. They found that the surface

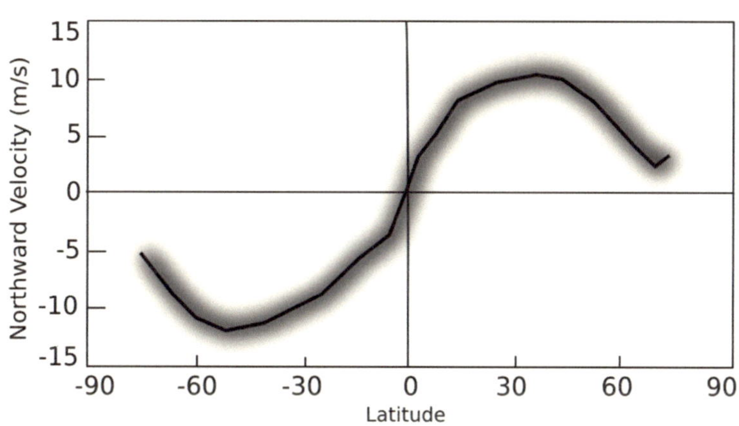

Figure 5.
Meridional flow velocity, approximate drawing from Hathaway and Rightmire [9].

meridional (following N-S meridians) flow velocities away from the equator are in the range of 0–15 m/s (**Figure 5**), negligible near to differential rotation (~170 m/s), granulation (~300 m/s), and supergranulation (~3000 m/s). The velocity is however responsible for the rate of polar magnetism reversion, thus the Sun cycle life span and its activity overall. Zhao and Kosovichev [10] studied the interior of a sunspot region using time-distance helioseismology [11] on a dataset of 512 uninterrupted dopplergrams at 1-minute cadence on August 7 and 8, 2000, from MDI, following the data preparation from Giles [12]. Vortical flows in the subphotospheric zone have been estimated through inverse modeling, leading to suggest that kinetic and magnetic helicity extends from surface to depth. Such connectivity could be a source of great energy buildup and enters into the making of solar flares.

4. Emerging flux characterization

Ilonidis et al. [13] found (from MDI data) some strong acoustic anomalies of 12–16 seconds as deep as 65 Mm, which became sunspots 1–2 days after detection with deep focus time-distance helioseismology (**Figure 6**). They ensured that measurements were made when magnetism was <300 G, otherwise masked out. They also found an emergence velocity of about 60 Mm in about 2 days, consistent with previously modeled velocities.

Shibata et al. [14] used the Solar Optical Telescope (SOT: hinode.nao.ac.jp/sot_e/) on-board Hinode [15] in the Ca II H band (396.85 nm) of the Broadband Filter Imager (BFI) corresponding to chromospheric heating. They studied upper chromosphere and lower corona magnetic reconnection producing high-speed jets being involved in lower coronal X-ray jets and Hα surges. They found that solar nanoflares (spicule jets) events happen and propagate further than gravity-bound expectations by slow-mode magneto-acoustic shocks or fast-mode nonlinear Alfvèn waves shocks. They argue that their actual findings are one order of magnitude less than the necessary energy to heat the corona. To add some energy, they suggest a multiscale presence of nanoflares in the upper atmosphere, which are not all visible with Hinode instruments.

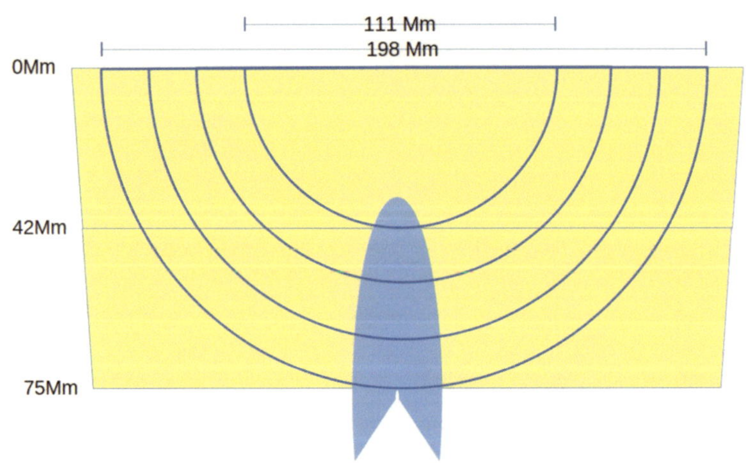

Figure 6.
Acoustic ray paths crossing an emerging flux from Ilonidis et al. [13].

Karoff and Kjeldsen [16] found that the background noise from granulation correlates with flare activity, and that increase in the background noise transfers more power into high-frequency modes via stochastic excitation, as observed in their experiment.

5. Penumbral characterization and co-processes

Ichimoto et al. [17] observed the dynamical processes of strongly magnetized plasma through the twisting motions of penumbral filaments. When magnetic reconnection happens, Katsukawa et al. [18] observed penumbral microjets in chromospheric layers above the penumbra (**Figure 7**). With less than 0.4 Mm width and a life span less than 1 minute, they are by all means small features around sunspots. They are similar to limb's spicules, dynamic fibrils in active regions and quiet Sun's mottles. They might be involved in the thermal source of the sunspot's coronal connectivity.

Scharmer et al. [19] used the Crisp Imaging SpectroPolarimeter (CRISP) on the Swedish 1 m Solar Telescope in La Palma, Spain, to study the neutral carbon line (C I) of downward Doppler velocities in the penumbral area of a sunspot. Evidence directs to (dark) downflows that can reach 8 km/s in the blueshift, with an estimated horizontal magnetic direction, while (bright) outward flows have about 50 degrees angle with the surface.

Howe et al. [20] compared Doppler surface measurements with global and local helioseismology (MDI, GONG). They show that subsurface shear fine-tuning just below the surface is hard to reconcile across methods in high accuracy levels while quite agreeing at equator for good accuracy levels. Zonal flow patterns agree largely across methods about rotation-rate residuals allowing for multiscale sources generating wave signals discrepancies.

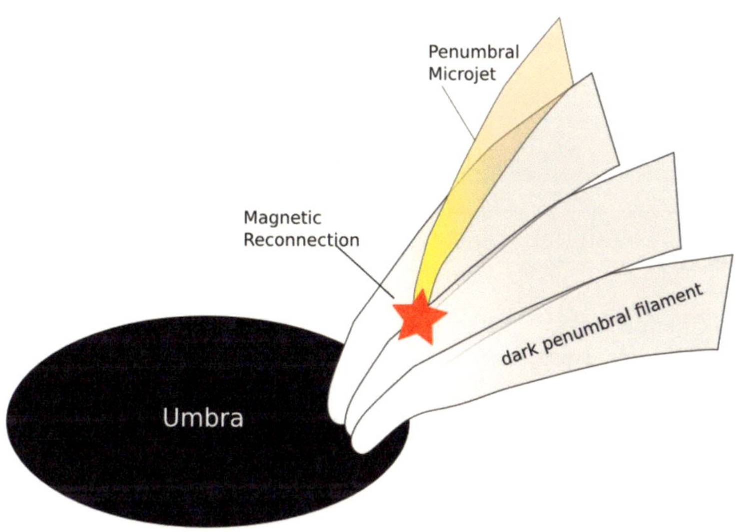

Figure 7.
Penumbral microjet and dark filaments [18].

6. Concluding remarks

Chromospheric and coronal events are strongly connected to magneto-convective dynamics in the photosphere, tachocline, and radiative layers below to some extents. There is a lot of unknown still on the particulars of the dynamics and interconnectivity across the Sun layers. Helioseismology and acoustic sciences have helped greatly in the recent years of discoveries.

The actual Solar Orbiter mission is not only returning remote sensing imagery of very fine resolution, as seen in the first part of this nonexhaustive chapter, but its dedicated sensors payload now starts generating significant amount of research on solar wind processes (i.e., [21] and the related special issue of A&A).

Author details

Yann-Henri Chemin
Joint Research Centre, Ispra, Italy

*Address all correspondence to: dr.yann.chemin@gmail.com

References

[1] Chaplin WJ, Kjeldsen H, Christensen-Dalsgaard J, Basu S, Miglio A, Appourchaux T, et al. Ensemble asteroseismology of solar-type stars with the NASA Kepler Mission. Science. 2011;**332**(6026):213-216

[2] Howe R. Dynamic variations at the base of the solar convection zone. Science. 2000;**287**(5462):2456-2460. DOI: 10.1126/science.287.5462.2456

[3] Amal'Skaya SI. The intergranular lanes in the quiet solar photosphere: Organized structures. Byulletin Solnechnye Dannye Akademie Nauk SSSR. 1990;**3**(March):87-94

[4] Puschmann KG, Ruiz Cobo B, Vázquez M, Bonet JA, Hanslmeier A. Time series of high resolution photospheric spectra in a quiet region of the Sun. II. Analysis of the variation of physical quantities of granular structures. AAP. 2005;**441**:1157-1169

[5] Phys.org. Zooming into the sun with Solar Orbiter by European Space Agency. Retrieved on the 25th of March, 2022. https://phys.org/news/2022-03-sun-solar-orbiter.html

[6] Thompson MJ, Toomre J, Anderson ER, Antia HM. Differential rotation and dynamics of the solar interior. Science. 1996;**272**(5266): 1300-1305

[7] Nandy D, Muñoz-Jaramillo A, Martens PCH. The unusual minimum of sunspot cycle 23 caused by meridional plasma flow variations. Nature. 2011;**471**(7336):80-82. DOI: 10.1038/nature09786

[8] McIntosh SW, De Pontieu B, Carlsson M, Hansteen V, Boerner P, Goossens M. Alfvenic waves with sufficient energy to power the quiet solar corona and fast solar wind. Nature. 2011;**475**(7357):477-480. DOI: 10.1038/nature10235

[9] Hathaway DH, Rightmire L. Variations in the Sun's Meridional flow over a solar cycle. Science. 2010;**327**(5971):1350-1352. DOI: 10.1126/science.1181990

[10] Zhao J, Kosovichev AG. Helioseismic observation of the structure and dynamics of a rotating sunspot beneath the solar surface. The Astrophysical Journal. 2003;**591**(1):446

[11] Duvall TL, Jeffferies SM, Harvey JW, Pomerantz MA. Time–Distance Helioseismology. Nature. 1993; **362**(6419):430-432

[12] Giles PM. Stanford University, 2000

[13] Ilonidis S, Zhao J, Kosovichev A. Detection of emerging sunspot regions in the solar interior. Science. 2011;**333**(6045):993-996. DOI: 10.1126/science.1206253

[14] Shibata K, Nakamura T, Matsumoto T, Otsuji K, Okamoto TJ, Nishizuka N, et al. Chromospheric anemone jets as evidence of ubiquitous reconnection. Science. 2007;**318**(5856):1591-1594. DOI: 10.1126/science.1146708

[15] Kosugi T, Matsuzaki K, Sakao T, Shimizu T, Sone Y, Tachikawa S, et al. The Hinode (Solar-B) Mission: An overview. Solar Physics. 2007;**243**(1):3-17

[16] Karoff C, Kjeldsen H. Evidence that solar flares drive global oscillations in the sun. The Astrophysical Journal Letters. 2008;**678**(1):L73

[17] Ichimoto K, Suematsu Y, Tsuneta S, Katsukawa Y, Shimizu T, Shine RA, et al. Twisting motions of sunspot penumbral filaments. Science. 2007;**318**(5856):1597-1599

[18] Katsukawa Y, Berger TE, Ichimoto K, Lites BW, Nagata S, Shimizu T, et al. Small-scale jetlike features in penumbral chromospheres. Science. 2007;**318**(5856):1594-1597. DOI: 10.1126/science.1146046

[19] Scharmer GB, Henriques VMJ, Kiselman D, Rodriguez. J d LC. Detection of convective downflows in a Sunspot Penumbra. Science. 2011;**333**(6040):316-319. DOI: 10.1126/science.1206429

[20] Howe R, Komm R, Hill F, Ulrich R, Haber DA, Hindman BW, et al. Large-scale zonal flows near the solar surface: A comparison of results from local and global helioseismology with direct doppler measurements. Solar Physics. 2006;**235**(1-2):1-15. DOI: 10.1007/s11207-006-0117-2

[21] Kieokaew R et al. Solar Orbiter observations of the Kelvin-Helmholtz waves in the solar wind. Astronomy & Astrophysics. 2021;**656**:A12

Chapter 8

Solar Proton Activity over the Solar Cycle 24 and Associated Space Radiation Doses

Wellen Rukundo

Abstract

The least number of proton events and ground-level enhancements was recorded in the solar cycle 24 which corresponds with the least smoothed sunspot number compared to the last three previous solar cycles. This was attributed to the weak sun's polar field and decreasing strength of the interplanetary magnetic field at the start of the solar cycle. The majority contribution to background radiation dose within our earth's atmosphere is galactic cosmic rays and trapped particles in the Van Allen Belts. However, solar proton events cause sudden spikes in radiation doses, and this depends on the fluence and energy spectra of the events. While these doses are least detected in the lower atmosphere, they have significant radiation damage to spacecraft electronic components and astronauts on long space missions and at higher atmospheric altitudes. Therefore, the prediction of such events and estimation of their effective radiation damage is an important consideration for planning long space missions and spacecraft design materials.

Keywords: energetic particles, galactic cosmic rays, fluence, energy spectra, radiation dose

1. Introduction

The sun's activity varies over the 11 years of a solar cycle with active and quiet phases. Over this period, several energetic particles are released into the interplanetary medium and accelerated towards the earth along the interplanetary magnetic field lines by several mechanisms discussed by Reames [1]. The sudden increase of energetic particle flux is detected by particle detectors on several satellites as solar proton events (SPEs). The proton events can also be detected by neutron monitors on ground stations as suppression of galactic cosmic ray (GCR) intensity (Furbish decrease), mostly observed on geomagnetic storm days [2]. For such proton events, they are defined as ground level enhancements (GLEs) and are associated with a 'hard' spectrum of energies >500 MeV [3]. Gradual SPEs are driven by CMEs and interplanetary shocks with a high proton to electron ratio while impulsive events are X-ray flashes associated with solar flares lasting for a few hours to about a day and with a high electron to proton ratio [1, 4, 5]. A few energetic protons from SPEs can escape the geomagnetic shielding and penetrate through the open magnetic field lines at the polar regions of the earth's atmosphere. These deposit their energy on spacecraft and satellite components at higher orbital altitudes causing space hazardous effects like spacecraft charging, single event upsets, and high radiation levels [6, 7]. The high radiation doses can also be detected by dosimeters onboard aircraft flying at high aviation altitudes mostly during intense proton and GLE events despite the atmospheric mass shielding [8–10]. GCRs consist of protons, alpha particles, and heavier nuclei which interact with the atmospheric constituents to produce secondary particles that can be observed at the ground level station by neutron and muon detectors. They have high energies ranging between 1 and 20 GeV, most of these energetic particles are deflected away by the magnetosphere and earth's magnetic field. The few penetrating particles contribute to background radiation levels within the earth's atmosphere. The important property of GCRs is the magnetic rigidity which defines their ability to resist bending in a magnetic field; this value varies from 0 at the poles to about 17 at the equatorial latitudes [11]. Therefore, the radiation doses due to GCRs are more enhanced at the poles than at the equator. Solar activity is a modulation factor for the propagation of GCRs and this is defined by an anti-correlation relationship between GCR intensity and sunspot number [12].

In this paper, we study the proton event, solar and cosmic ray activity during the solar cycle 24 and relate them to space radiation doses. We study and discuss the effect of particle fluence and energy spectra on dose contribution and some of the effects of high space radiation doses. This is an important consideration for planning long space missions and designing appropriate shielding materials for spacecraft manufacturers.

2. Solar proton events and data

The energetic proton events are detected by several detectors onboard the GOES spacecraft at Geosynchronous orbit [13] and the list of proton events is available at NOAA SWPC (https://ngdc.noaa.gov/stp/space-weather/interplanetary-data/solar-proton-events/SEP%20page%20code.html). The integrated energies for 5-min averages >10 MeV, measured in particle flux units (pfu) defined the proton flux. The start of the SPE was considered when the first three consecutive fluxes were ≥10 pfu and the end was the last time when the flux was greater than or equal to 10 pfu. The last three events in 2017 were picked from the list of Solar Proton Event Archive by European Space Agency (ESA) (https://space-env.esa.int/noaa-solar-proton-event-archive/). Each proton event was verified in the GLE list available at Neutron Monitor Database Event Search Tool (NEST) (https://www.nmdb.eu/nest/gle_list.php) to confirm if it's associated with the GLE.

2.1 SPE relationship with solar and GCR activity

SPEs are accompanied by shock and flare components and their observation depends on the connectivity of the observer to the flare site. The X-ray flares are used to classify the size of the events with X and M-class being the most powerful and occurring more frequently during the active phase of the solar cycle. SPEs occurrence peaks during the solar maxima with frequent spikes of proton flux occurring due to super active regions associated with large successive eruptions. Out of the total 42 recorded SPEs, only 6 were associated with weak C-class flare eruption while the rest were associated with X and M-class flares. The C-class flares associated SPEs occurred mostly during the ascending phase of the solar cycle. Most energetic events with pfu exceeding

100 pfu occurred during the solar maxima/active periods (86% of the events with >100 pfu occurred from 2012 to 2015). **Figure 1** shows the bar graph of the yearly number of SPEs from 2010 to 2017 with the yearly averaged sunspot number. The number of SPEs for the four consecutive solar cycles were 59, 72, 79, and 42 for solar cycles 21, 22, 23 and 24 respectively (https://ngdc.noaa.gov/stp/space-weather/interplanetary-data/solar-proton-events/SEP%20page%20code.html) while the maximum smoothed sunspot number is 232.9, 212.5, 180.3, and 81.8 for the four solar cycles respectively (https://wwwbis.sidc.be/silso/datafiles) [14]. The sharp decline of sunspot number from solar cycle 23–24 corresponds with the least number of SPEs in solar cycle 24. The occurrence rate of GLE events follows a similar pattern with 18%, 20%, 22%, and 2% of the total GLE events (72) for the 4 solar cycles respectively. The least numbers of SPEs and GLEs were recorded in the solar cycle 24 with the least maximum smoothed sunspot number. Solar cycle 24 has been ranked as the fourth weakest of the 24 solar cycles since 1755 and the weakest in 100 years (https://www.weather.gov/news/201509-solar-cycle) [15]. This was caused by the impaired growth of the polar field. The simulation of the solar surface field by Jiang and Schüssler [16] showed emerging low latitude bipolar regions with an opposite orientation of the magnetic polarities in the north–south direction. This was the cause of growth impairment of the polar field and hence resulted in a weak magnetic field throughout the solar cycle 24. **Figure 2** shows the variation of sunspot, proton flux at energy levels of >10, >30 and > 60, X-ray intensity, and GCRs intensity recorded by Oulu neutron monitor from 2010 to 2017. The intensity of GCRs recorded by neutron monitors shows an anti-correlation relationship with the rate of particle flux entering the earth's atmosphere and sunspot number [12, 17].

2.2 Ground level enhancements

GLE events have more concerns with space radiation due to a harder spectrum and large energy intensity. Mewaldt et al. [18] estimated an average power-law index of about 3.18 at energies of >40 MeV and energy intensity of >50 MeV for GLE events in solar cycle 23. He also found out that these events exhibited larger ratios of Fe/O ratios and 3He-rich element composition. **Table 1** shows the detailed properties of the two proton events associated with GLE that occurred during the solar cycle 24. The properties of the GLE events were obtained from Gopalswamy et al. [19]—GLE 71 and

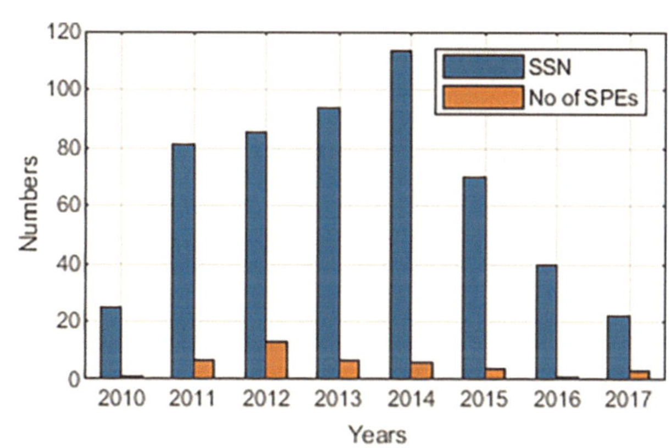

Figure 1.
A bar graph showing the yearly averaged sunspot number and yearly number of SPEs from 2010 to 2017.

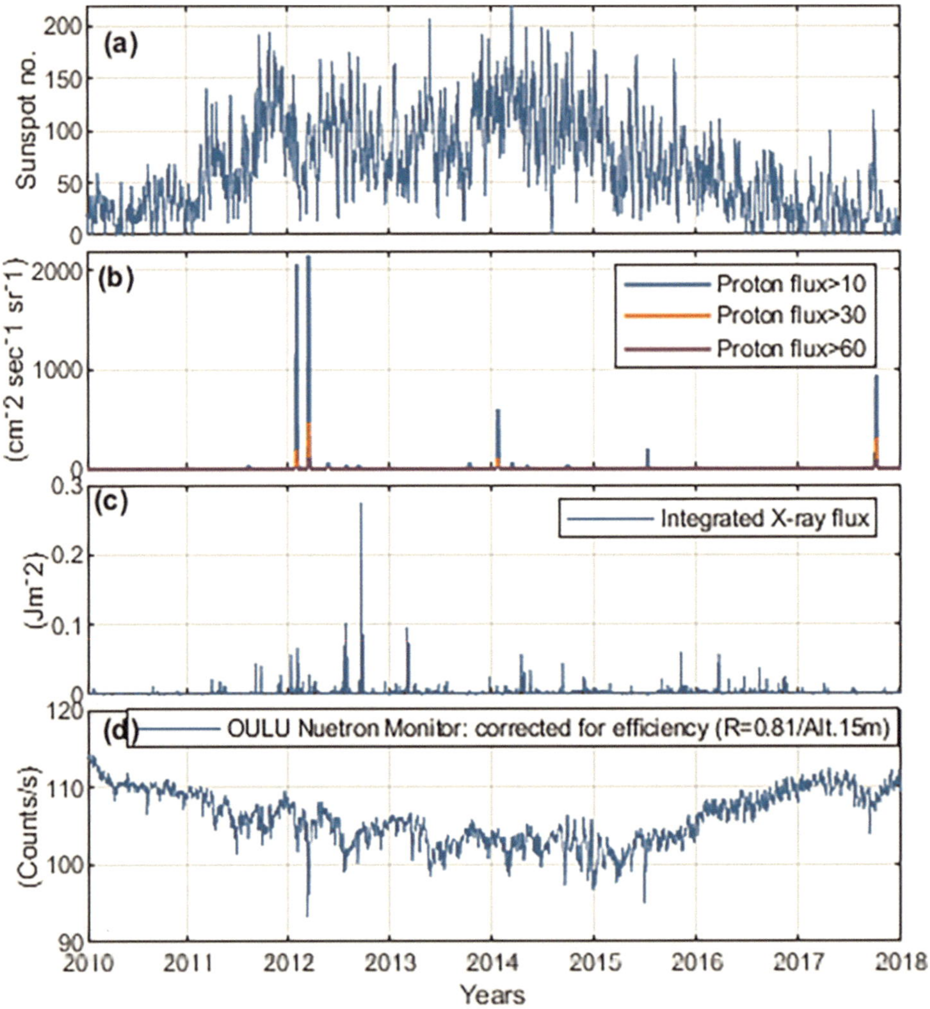

Figure 2.
The variation of sunspot number (a), proton flux at three energy levels (b), integrated X-ray flux (c), and cosmic ray intensity recorded at Oulu neutron monitor station (d) over solar cycle 24.

Cohen and Mewaldt [20] and Gopalswamy et al. [21]—GLE 72. It's observed that the GLE events are associated with high CME shock speeds, powerful X-ray flares and originated from the western hemisphere. This provides a better magnetic connection between the sun's active region and the observer along the sun-earth line [1, 22, 23].

3. Radiation doses within the near-earth space environment from SPEs

Energetic particles driven by CME shocks and flares and GCRs that escape the magnetospheric shielding through the polar regions get trapped by the earth's magnetic field in the Van Allen Belts (VAB). The outer belts consist of energetic electrons with energies up to ≈10 MeV while the inner belt consists of electrons and protons with energies ranging from KeV to ≈10 MeV [24, 25]. Over the South Atlantic Anomaly (SAA) which extends from South America to West Africa, the VAB dips

#	GLE 71	GLE 72
Start time	2012 05/170210	2017 09/101625
End time	2012 05/170430	2017 09/160455
>10 MeV maximum pfu	255	844
Region	11,476	12,673
Location	N11W76	S14W74
Flare class	M5.1	X8.3
CME speed (km/s)	1582	3163
Flare time	01:25	15:35
CME time	01:48	16:00
Type II	Yes	Yes
Max energy	>700	>700
Shock height	3.06 Rs	3.92 Rs
% enhancement	18.6	6

Table 1.
A table showing the properties of proton events associated with GLE in the solar cycle 24.

down to about 200 km into the upper atmospheric region; the increased energetic particle flux in this region contributes the largest radiation dose during low inclination flights and space missions. The secondary particle showers from the ionization of GCRs contribute mostly to the background radiation levels within our earth's atmosphere throughout the solar cycle. Several spacecraft and satellites launched into different earth orbits and altitudes have inbuilt particle detectors for detecting the energetic particle flux. The energy deposited by these particles is evaluated by models developed using either radiation transport equations or Monte-Carlo calculations to calculate the amount of absorbed energy as radiation dose to the spacecraft.

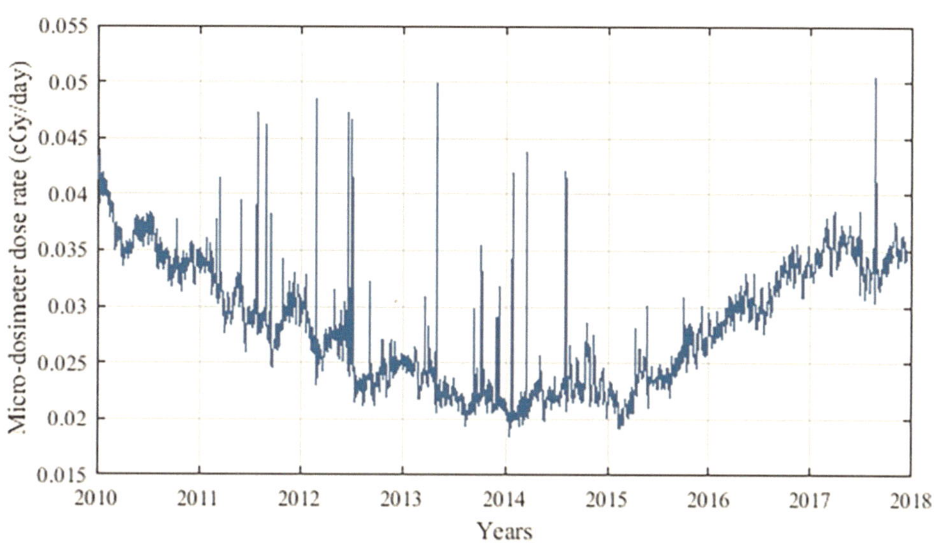

Figure 3.
Dose recorded by a micro-dosimeter on CRaTER from 2010 to 2017.

Figure 3 shows the dose recorded by a micro-dosimeter from 2009 to 2017 on CRaTER (https://prediccs.sr.unh.edu//data/craterProducts/doserates/data/2017365/doserates_micro_2017365_alldays_allevents.txt). The sudden increase/spikes in dose were due to protons events which were more frequent during the solar maxima periods. The CRaTER is an experiment on the Lunar Reconnaissance Orbiter (LRO), a NASA spacecraft orbiting the moon. CRaTER characterizes the radiation environment with measurement of effects of ionizing energy loss on silicon solid detectors (three pairs of thin and thick silicon detectors and a micro dosimeter) due to penetrating energetic particles and GCRs. The micro dosimeter is on the analog electronics board with the aluminum shield of a thickness of 2.28 g/cm^2 facing space [26].

Figure 4 shows the correlation relationship (*R*) between dose from 2010 to 2017 and GCR intensity and sunspot number. GCR intensity was got from Oulu Neutron monitor in the NEST database (https://www.nmdb.eu/nest/) and sunspot number was got Omniweb database (https://omniweb.gsfc.nasa.gov/form/dx1.html). **Figure 4a** clearly shows that there is a high positive correlation between the dose and the GCR intensity while **Figure 4b** shows a significant negative correlation between dose and sunspot number. Solar activity modulates GCRs with 90% of the GCRs filtered out by heliospheric solar wind [27]. This effect decreases with solar wind pressure decrease and this results in change in the controlling heliospheric transport processes. The heliospheric transport processes (diffusion, convection, and adiabatic acceleration) are dominated by cosmic rays during solar minima however this shifts to solar wind dominated transport processes during the solar maxima. This suppresses GCRs propagation, hence a decrease in count rate is observed on the ground neutron detectors drinng solar maxima. The correlation relationship between dose and GCRs and SSN conforms to an anti-corelation relation observed between GCRs and sunspot in **Figure 2**, this has been studied by a number of researchers [28, 29]. It should be noted that the modulation of GCR intensity does not only depend on solar activity but also on interplanetary magnetic field which further modifies the transport processes [29]. We can observe from **Figure 4b** that the sunspot number has a less effect on space radiation but only a factor that describes solar activity variation and therefore has a positive relationship with occurrence of proton events.

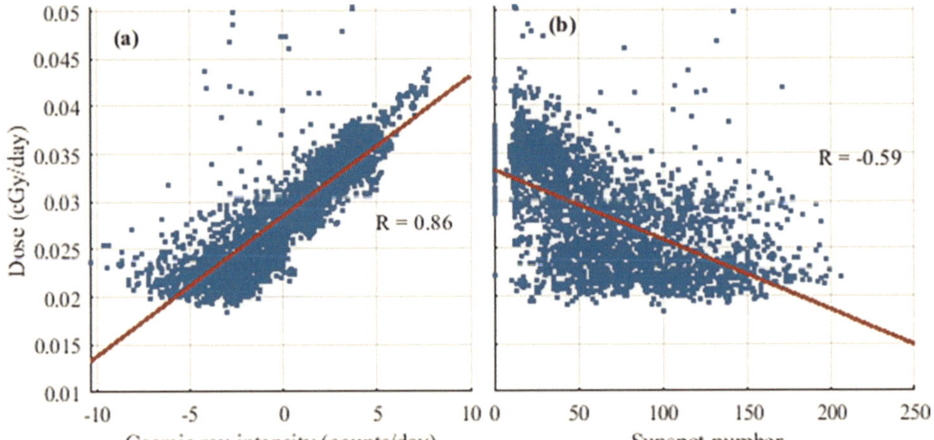

Figure 4.
Correlation plot between micro-dosimeter reading from 2010 to 2017 and cosmic ray intensity (a) and sunspot number (b).

3.1 Effects of radiation doses

Radiation damage effects are discussed extensively by several scientists including [7, 9, 30] and references therein. In transistors, the accumulation of total ionizing dose (TID) builds up positive charges on the gate oxide of field-effect transistors causing threshold shifts and off-state leakage currents. At very high dose rates, shifts become large enough to exceed the threshold, this results in defective and poor responsiveness of the critical spacecraft component [30]. In semiconductors, the deposited energy by non-ionization processes is quantified as the displacement damage dose (DDD). The DDD displaces electrons from their initial positions forming vacancies and interstitials—Frenkel pairs. The localized groupings of Frenkel pairs result in the formation of material defects which alters the material properties [6, 31]. This affects satellite components mostly utilizing semiconductors; the most observable effect is the loss of maximum power output due to the degradation of solar cells. Hands et al. [30] evaluated degradation performance on solar cells by evaluating cumulative damage effects of TID and DDD on a Galileo satellite and concluded that the most damage effects of extreme space weather come from trapped electrons rather than solar protons.

Astronauts on long space missions may be affected by cumulative radiation effects and exposed to short periods of high radiation levels during proton events. The radiation effects on humans are either deterministic resulting from immediate effects due to high sudden dose levels or stochastic associated with long-term effects of abnormal tissue growth and cancer arising from cumulative doses. Previous studies using space environment radiation models have calculated an average total body dose of ≈ 0.5–1.5 μGy/min for space missions which may increase in case of SPE occurrence [32, 33]. In comparison, the dose rates are $\approx 10^8 - 10^{10}$ times lower than the medical exposure dose rates (8, 4 and 2 mGy for abdominal computed tomography, mammography, and radiotherapy sessions respectively). However, radiation-induced cataracts, DNA damage, cancer risks, cell damage, chromatin decondensation are possible biological effects from repeated doses and nonlinear low dose effects [33]. The estimation of biological space radiation damage is more complex attributed to the probabilistic nature of SPE occurrence, very low doses, individual susceptibility and tissue response to radiation, quantification of secondary radiation, and limited dose data for spacecraft crews on long space missions to develop more representative models for radiation effects.

Particle fluxes of secondary cosmic radiation at 18 and 12 km flight altitude are about 500 and 300 times greater than at sea level respectively [34]. These are the main sources of background radiation at aviation flight altitudes. However, very large and intense SPEs and GLEs may increase the radiation exposure levels to exceed the recommended threshold safety levels (effective dose limits of 20 mSv/year averaged over five years for radiation workers and 1 mSv/year for the public) by the international radiation protection community and the International Civil Aviation Organization (ICAO) [35]. Dosimeter measurements and model simulations have shown an increase in radiation levels during specific proton and GLE events, for example, radiation dose levels on Lufthansa flights between Munich and Chicago rose from 3.4 to 5.7 μSv/hr. during GLE 65 [36] while on Qantas 747 flight from Los Angeles to New York City, the radiation levels increased from 3.4 to 4.7 μSv/hr. during the GLE 66 [37], simulated effective dose rates at 12 km altitude were 5.8 μSv/hr. for GLE 70 and 4.5 μSv/hr. for non-GLE SPE (9th November 2000) [8]. Mishev and Usoskin [38] estimated about 100 μSv for a polar flight at 12.5 km altitude for 3.5 hours from during GLE 72 using rigidity spectrum. While the modeled and simulated doses obtained may not be the actual effective radiation doses to the aircrew but it's indicative of a possibility of radiation hazard on the aviation industry associated with SPEs. A survey on radiation exposure to Canadian pilots showed an annual dose of about 3 mSv [39]. However, the estimation of total effective radiation doses to the aircrew requires consideration of flight path, time of flight, flight altitude, the position of the aircrew within the aircraft, and material component of the aircraft (shielding design). The International Commission on Radiation Protection (ICRP) recommendations in 2007 recommended that the aircrew be considered as occupationally exposed workers with the implementation of practical regulatory measures including monitoring programs and individual dose assessment for radiation protection by aviation operators [40]. Several warning systems of aviation dosimetry [10, 41, 42] have been developed in addition to S-scale provided by NOAA SWPC alerts the aviation operators to initiate the safety protocols which include rerouting, flying at low altitudes, avoiding polar regions, and flight delay. However, they are challenged by the probabilistic occurrence of SPEs and the description of the broad energy spectrum from low to high energies [10].

3.2 Quantification of space radiation doses from SPEs

Energetic particles incident on spacecraft and satellite materials deposits energy by ionization process, these particles include proton, trapped protons, and electrons in the VAB, secondary photons, and GCRs. The energy absorbed is measured as TID of which cumulative effects can result in device failure of critical spacecraft electronics and biological damage to astronauts. TID is quantified by absorbed dose (SI unit called rad, that is 1 gray (Gy) = 100 rads = 1 J/kg). The absorbed dose (D) is a function of fluence spectrum at different energy levels (∅) and the mass stopping power of the material ($dE/\rho dx$) through which the incident particle penetrates (Eq. (1)); this is dependent on the orbit altitude, time taken in the orbit, and spacecraft/satellite orientation [43].

$$D = \varnothing \frac{dE}{\rho dx}\left(1.6 \times 10^{-10}\right)\text{ Gy} \tag{1}$$

The Bethe-Bloch formula [44] defines stopping power using relativistic quantum mechanics (Eq. (2)):

$$\frac{dE}{dx} = \frac{4\pi z^2 k_o^2 e^4}{mv^2}\left[\ln \frac{2mv^2}{I} - \ln\left(1 - \frac{v^2}{c^2}\right) - \frac{v^2}{c^2}\right] \tag{2}$$

where z is the atomic number of the heavy particle, e is the magnitude of the electron charge, m is the electron rest mass, c is the speed of light, I is the mean excitation energy of the medium, v is the velocity of the particle and k_o is the Boltzmann constant.

3.3 Fluence and energy spectra dose relationship

Space Environments and Effects section at the European Space Agency maintains an archive of SPEs with plots and fit data for event fluence and radiation effects (https://space-env.esa.int/noaa-solar-proton-event-archive/) from 1997 up to date. The event selection is based on the NOAA/GOES-p_ 5 min averaged >10 MeV p + flux exceeds 2.0 p+/cm^2/s/sr and ends when the flux returns to below 1 p+/cm^2/s/sr. The estimation of radiation effects is done using the best fit from a comparison of three forms of fit that is exponential in power rigidity $\left(A * e^{-Rm/B}\right)$, exponential in energy $\left(A * e^{-E/B}\right)$ and power law $\left(A * E^{B}\right)$, where A and B are constants, Rm is magnetic rigidity and E is energy. **Figure 5a–f** shows the correlation relationship (R) of the SPEs integrated fluence spectra from 2010 to 2017 at different proton energies of 1, 5, 10, 30, 50 and 100 MeV with the dose behind 0.05 mm Al shield in rads. The R values show that the integrated fluence spectrum at all proton energies has a positive correlation with the magnitude of the dose. However, the integrated fluence at low energies from 1 to 30 MeV is more correlated to dose with high R values compared to higher

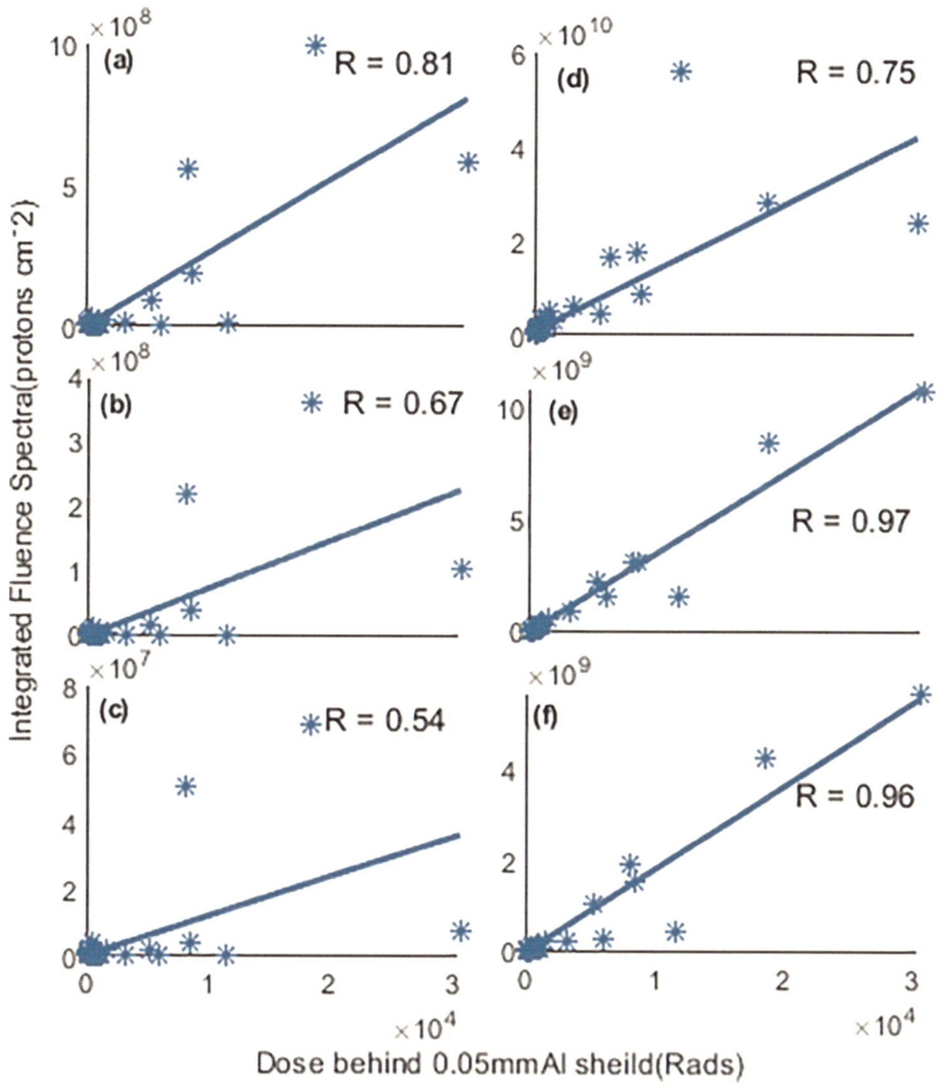

Figure 5.
Scatter graphs with line of best fit showing correlation relationship between integrated fluence spectrum and dose at different energy levels; (a) 30 MeV, (b) 50 MeV, (c) 100 MeV, (d) are 1 MeV, (e) 5 MeV and (f) 10 MeV.

energies from 50 to 100 MeV. This observation is a manifestation of the contribution of slowed-down proton spectra to dose build-up on a material.

4. Discussion

4.1 Fluence dose relationship

Dose distributed over a material results from a specific number of particles incident on a given material surface. Using the simplified equation of dose, $D = \varnothing_p(E)S_\mathrm{p}(E)$, where S_p corresponds to non-ionizing energy loss (NIEL) for DDD and stopping power for TID. The incident primary fluence interacts with the material through soft and hard knock-on collisions resulting in slowed down secondary charged particles. This develops a charged particle equilibrium with the absorption of the secondary fluence as the radiation energy escapes [45]. There exists a linear relationship existing between proton damage coefficients and determined NIEL for the case of GaAs cells [46]. The observable relationship results from the DDD increasing with the increase in proton fluence. Therefore, proton fluence at any energy level is an input to calculating degradation curves and or characteristic curves from which relative proton damage coefficients can be determined.

4.2 Energy spectra dose relationship

While the less energetic protons incident on the material is shielded off, the highly energetic protons (>100 MeV) are slowed down by the shielding of the material to low energies. The slowed-down protons initiate atomic and molecular collision; this process needs a build-up from low energies to exceed the threshold energy for atomic displacement to occur. Once the threshold energy is exceeded, the material properties start to degrade with absorption and particle production. Comparing the rate of energy loss, the NIEL peaks faster at much lower energies than the stopping power. While different materials have different threshold energies for atomic displacement, low proton energies (<1 MeV) are enough to build the threshold energies for atomic displacement for most shielding materials and this plays the largest contribution to damage production. The defining functional form of the energy spectrum below 1 MeV behaves as a reciprocal of stopping power calculated using the continuous slowing down approximation (CSDA) theory and is independent of the shield thickness. Theoretically, this is the range of the proton (Ra) which defines the mean distance a proton travels in the matter before it stops (Eq. (3)) below adopted from [6].

$$\mathrm{Ra} = \int_{E_\mathrm{min}}^{E_\mathrm{max}} -\frac{\rho \mathrm{d}x}{\mathrm{d}E}\,\mathrm{d}E + \mathrm{Ra}(E_\mathrm{min}) \tag{3}$$

Where $\mathrm{Ra}(E_\mathrm{min})$ is the measured range at minimum energy (E_min) always taken at 1 MeV due to large enough data available for any proton event at the energy level.

Messenger et al. [47] found that 90% of the total dose to the shielded device was a contribution of relatively low proton energies up to about 12 MeV and only 10% was from proton energies from 12 to 500 MeV for an event on 19th October 1989. Extremely large SPEs associated with GLEs rarely occur, these contain high proton energies extending to GeV. The energy spectrum from such events is known to be 'hard' and it can penetrate deeper thicknesses of the shielding material. Mertens and Slaba [48] calculated the cumulative dose from a set of 65 historical SPEs associated with GLEs using the power-law function and noted that 25% of the peak total dose was contributed from energies >500 MeV corresponding to a spectrally hard SPE of February 1956. The double power law of the band distribution function has been recommended to best describe the energy spectrum of a GLE event over the broad energy ranges [18, 49]. Also, Xapsos et al. [50] found that the Weibull distribution describes the proton event energy spectra for the smallest values and is appropriate on broad energy ranges.

5. Summary

The occurrence of solar proton and GLEs events follow solar cycle described by sunspot number. A least number of SPEs were recorded in solar cycle 24 with the least smoothed sunspot number compared to the previous three solar cycles. Every proton energy starting from the lowest to the highest energy level contributes to radiation dose and the slowed-down protons at low energies are critical in the assessment of space radiation damage. I also noted that the shield of a material is an important component to consider when evaluating proton damage effects. It's important to use the most appropriate distribution function that can describe the energy spectrum from low to high energy levels for any SPE event. However, this is a challenge with the GLE associated SPEs where there is an underestimation at high energies of GeV by most distribution functions. A further study regarding the probabilistic occurrence of SPEs, with the associated properties of solar flares, CMEs, GLE events, and energy spectrum is necessary, this is important to improve modeling and prediction capabilities of such events.

Acknowledgements

I acknowledge several data providers which include NOAA SWPC, ESA, NEST, SILSO, Omniweb, and NASA.

Author details

Wellen Rukundo
Space Environment and Research Laboratory, Institute of Basic and Applied Science, Egypt Japan University of Science and Technology (E-JUST), Alexandria, Egypt

*Address all correspondence to: wellen.rukundo@ejust.edu.eg

References

[1] Reames DV. Particle accelaration at the sun and in the heliosphere. Space Science Reviews. 1999;**90**(3/4):413-491

[2] Kang J, Oh S, Yi Y, Kim Y. Forbush decreases observed by Daejeon neutron monitor. Advances in Space Research. 2016;**57**(3):912-918

[3] Kuwabara T, Bieber JW, Clem J, Evenson P, Pyle R, Munakata K, et al. Real-time cosmic ray monitoring system for space weather. Space Weather. 2006; **4**(8):10

[4] Shimazu H. Solar proton event and proton propagation in the earth's magnetosphere. Journal of the National Institute of Information and Communications Technology. 2009;**56**: 191-198

[5] Hu S. Solar particle events and radiation exposure in space. The Health Risks of Extraterrestrial Environments. 2017;**3**:15

[6] Nwankwo VU, Jibiri NN, Kio MT. The impact of space radiation environment on satellites operation in near-earth space. In: Demyanov V, Becedas J, editors. Satellites Missions and Technologies for Geosciences. London: IntechOpen; 2020. p. 17

[7] Zheng Y, Ganushkina NY, Jiggens P, Jun I, Meier M, Minow JI, et al. Space radiation and plasma effects on satellites and aviation: Quantities and metrics for tracking performance of space weather environment models. Space Weather. 2019;**17**(10):1384-1403

[8] Kataoka R, Nakagawa Y, Sato T. Radiation dose of aircrews during a solar proton event without ground-level enhancement. Annales of Geophysics. 2015;**33**:75-78

[9] Dyer C. Radiation effects on spacecraft & aircraft. European Space Agency, (Special Publication) ESA. 2002;**477**:505-512

[10] Meier MM, Matthiä D. A space weather index for the radiation field at aviation altitudes. Journal of Space Weather and Space Climate. 2014;**4**(A13):5

[11] Bütikofer R. Cosmic ray particle transport in the Earth's magnetosphere. In: Shore SN, editor. Solar Particle Radiation Storms Forecasting and Analysis. Cham: Springer; 2017. pp. 79-94

[12] Ross E, Chaplin WJ. The behaviour of galactic cosmic-ray intensity during solar activity cycle 24. Solar Physics. 2019;**294**(1):17

[13] Sauer HH. GOES observations of energetic protons to E > 685 MeV: Description and data comparison. In: Leahy DA, Hicks RB, Venkatesan D, editors. 23rd International Cosmic Ray Conference, 19-30 July 1993, Singapore. Alberta, Canada: University of Calgary; 1993

[14] SILSO World Data Center. The International Sunspot Number. International Sunspot Number Monthly Bulletin and online catalogue. 1976–2018

[15] National Oceanic and Atmospheric Administration. Hello Solar Cycle 25 [Internet]. 2020. Available from: https://www.weather.gov/news/201509-solar-cycle [Accessed 23 October 2021]

[16] Jiang J, Cameron RH, Schüssler M. The cause of the weak solar cycle 24. The Astrophysical Journal. 2015;**808**(L28):6

[17] Gupta M, Mishra VK, Mishra AP. Long-term modulation of cosmic ray

intensity in relation to sunspot numbers and tilt angle. Journal of Astrophysics and Astronomy. 2006;**27**(4):455-464

[18] Mewaldt RA, Looper MD, Cohen CMS, Haggerty DK, Labrador AW, Leske RA, et al. Energy spectra, composition, and other properties of ground-level events during solar cycle 23. Space Science Reviews. 2012;**171**:97-120

[19] Gopalswamy N, Xie H, Yashiro S, Akiyama S, Mäkelä P, Usoskin IG. Properties of ground level enhancement events and the associated solar eruptions during solar cycle 23. Space Science Reviews. 2012;**171**(1):23-60

[20] Cohen CMS, Mewaldt RA. The ground-level enhancement event of September 2017 and other large solar energetic particle events of cycle 24. Space Weather. 2018;**16**(10): 1616-1623

[21] Gopalswamy N, Yashiro S, Mäkelä P, Xie H, Akiyama S, Monstein C. Extreme kinematics of the 2017 September 10 solar eruption and the spectral characteristics of the associated energetic particles. Astrophysical Journal Letters. 2018;**863**(L39):6

[22] Reames DV. The two sources of solar energetic particles. Space Science Reviews. 2013;**175**(1–4):53-92

[23] Gopalswamy N, Makela P. Properties of DH type II radio bursts and their space weather implications. In: URSI Asia-Pacific Radio Science Conference, New Delhi; 29 July 2019

[24] Ganushkina NY, Dandouras I, Shprits YY, Cao J. Locations of boundaries of outer and inner radiation belts as observed by cluster and double star. Journal of Geophysical Research. 2011;**116**(A9):18

[25] Gusev AA, Pugacheva GI, Jayanthi UB, Schuch N. Modeling of low-altitude quasi-trapped proton fluxes at the equatorial inner magnetosphere. Brazilian Journal of Physics. 2003;**33**(4):767-774

[26] Spence HE, Case AW, Golightly MJ, Heine T, Larsen BA, Blake JB, et al. CRaTER: The cosmic ray telescope for the effects of radiation experiment on the lunar reconnaissance orbiter Mission. Space Science Reviews. 2010;**150**(1–4): 243-284

[27] Florinski V. Galactic cosmic ray transport in the global heliosphere. Journal of Geophysical Research. 2003; **108**(A6):16

[28] Utomo YS. Correlation analysis of solar constant, solar activity and cosmic ray. In: 2nd International Symposium on Frontier of Applied Physics (ISFAP 2016), Jakarta, Indonesia; 3-5 October 2016

[29] Aslam OPM, Badruddin. Solar modulation of cosmic rays during the declining and minimum phases of solar cycle 23. Solar Physics. 2012;**279**:269-288

[30] Hands ADP, Ryden KA, Meredith NP, Glauert SA, Horne RB. Radiation effects on satellites during extreme space weather events. Space Weather. 2018;**16**(9):1216-1226

[31] Srour JR, Marshall CJ, Marshall PW. Review of displacement damage effects in silicon devices. IEEE Transactions on Nuclear Science. 2003;**50**(3):653-670

[32] Cucinotta FA, Durante M. Cancer risk from exposure to galactic cosmic rays: Implications for space exploration by human beings. Lancet Oncology. 2006;7(5):431-435

[33] Maalouf M, Durante M, Foray N. Biological effects of space radiation on

human cells: History, advances and outcomes. Journal of Radiation Research. 2011;**52**:126-146

[34] Hubert G, Aubry S, Clemente JA. Impact of ground-level enhancement (GLE) solar events on soft error rate for avionics. IEEE Transactions on Aerospace and Electronic Systems. 2020; **56**:3674-3684

[35] National Council on Radiation Protection and Measurements. NCRP Report No. 160, Ionizing Radiation Exposure of the Population of the United States. National Council on Radiation Protection and Measurements, Bethesda, Maryland; 2009

[36] Beck P, Latocha M, Stehno G, Rollet S. TEPC reference measurements at aircraft altitudes during a solar storm. Advances in Space Research. 2005;**36**(9): 1627-1633

[37] Getley IL. Observation of solar particle event on board a commercial flight from Los Angeles to New York on 29 October 2003. Space Weather. 2004;**2**:7

[38] Mishev AL, Usoskin IG. Assessment of the radiation environment at commercial jet-flight altitudes during GLE 72 on 10 September 2017 using neutron monitor data. Space Weather. 2018;**16**:1921-1929

[39] Bennett LGI, Lewis BJ, Bennett BH, Mccall MJ, Bean M, Doré L, et al. A survey of the cosmic radiation exposure of air Canada pilots during maximum galactic radiation conditions in 2009. Radiation Measurements. 2013;**49**: 103-108

[40] International Commission on Radiological Protection. ICRP. P103: The 2007 Recommendations of the International Commission on Radiological Protection. Amsterdam, Netherlands: Elsevier; 2007

[41] Sato T, Kataoka R, Yasuda H, Yashiro S, Kuwabara T, Shiota D, et al. Air shower simulation for WASAVIES: Warning system for aviation exposure to solar energetic particles. Radiation Protection Dosimetry. 2014;**161**(1–4): 274-278

[42] Latocha M, Beck P, Rollet S. AVIDOS—A software package for European accredited aviation dosimetry. Radiation Protection Dosimetry. 2009; **136**(4):286-290

[43] Jibiri NN, Nwankwo VUJ, Kio M. Determination of the stopping power and failure-time of spacecraft components due to proton interaction using GOES 11 acquisition data. International Journal of Engineering Science and Technology (IJEST). 2013; **3**(8):532-6542

[44] Sigmund P. General aspects and stopping of swift point charges. In: Particle Penetration and Radiation Effects. Berlin: Springer Berlin Heidelberg; 2006. p. 437 XXIV

[45] Alpen EL. Quantities, units, and definitions. In: Alpen EL, editor. Radiation Biophysics. 2nd ed. San Francisco, California: Academic Press; 1998. pp. 1-10

[46] Messenger SR, Summers GP, Burke EA, Walters RJ, Xapsos MA. Modeling solar cell degradation in space comparison of the NRL displacement damage dose and the JPL equivalent fluence approaches, Progress in Photovoltaics: Research and Applications. 2001;**9**(2):103-121

[47] Messenger SR, Xapsos MA, Burke EA, Walters RJ, Summers GP. Proton displacement damage and

ionizing dose for shielded devices in space. IEEE Transactions on Nuclear Science. 1997;**44**(6):2169-2173

[48] Mertens CJ, Slaba TC. Characterization of solar energetic particle radiation dose to astronaut crew on deep-space exploration missions. Space Weather. 2019;**17**(12):1650-1658

[49] Atwell W, Tylka A, Dietrich W, Badavi F, Rojdev K. Spectral Analyses and Radiation Exposures from Several Ground-Level Enhancement (GLE) Solar Proton Events: A Comparison of Methodologies. In: 41st International Conference on Environmental Systems, International Conference on Environmental Systems (ICES), Portland, Oregon; 17-21 July 2011

[50] Xapsos MA, Barth JL, Stassinopoulos EG, Messenger SR, Walters RJ, Summers GP, et al. Characterizing solar proton energy spectra for radiation effects applications. IEEE Transactions on Nuclear Science. **47**(6):2218-2223

Chapter 9

Ramjet Acceleration of Microscopic Black Holes within Stellar Material

Mikhail V. Shubov

Abstract

In this work, we present a case that Microscopic Black Holes (MBH) of mass 10^{16} kg–3×10^{19} kg experience acceleration as they move within stellar material at low velocities. The accelerating forces are caused by the fact that an MBH moving through stellar material leaves a trail of hot rarefied gas. The rarefied gas behind an MBH exerts a lower gravitational force on the MBH than the dense gas in front of it. The accelerating forces exceed the gravitational drag forces when MBH moves at Mach number $\mathcal{M} < \mathcal{M}_0 < 1$. The equilibrium Mach number $\mathcal{M}_0$ depends on MBH mass and stellar material characteristics. Our calculations open the possibility of MBH orbiting within stars including the Sun at Mach number $\mathcal{M}_0$. At the end of this work, we list some unresolved problems which result from our calculations.

Keywords: primordial black holes, microscopic black holes, ramjet acceleration, accretion, intrastellar orbits

1. Introduction

In the research presented in the works [1–3], it has been suggested that Primordial Black Holes make up a significant fraction of dark matter. Microscopic Black Holes (MBH) can also be formed within stars by coalescence of dark matter composed of weakly interacting massive particles [4, 5]. According to the plot in ([3], p. 14), considerations other than stellar capture constrain the masses of MBH as a dark matter to the range of 10^{16} kg–5×10^{21} kg.

Up to now, researchers believed that all MBH captured by a star would be slowed down within stellar material until they settle in the stellar center [1, 2]. In the present work, we explore the possibility of MBH accelerating during their passage through stellar matter at low Mach numbers. As MBH passes through matter, it accretes material at a rate we denote $\dot{M}$. Some of the mass accreted by MBH is turned into energy. This energy escapes the MBH in the form of protons and gamma rays. These rays heat the surrounding material, causing its rarefaction. The rarefied material behind the moving MBH exerts a lower gravitational pull on the MBH than the dense material in front of it. *Moving MBH experiences a net forward force.* This force is called **MBH ramjet** force. The effect is illustrated in **Figure 1**.

The conditions under which MBH accelerates within the stellar material are derived in this work. In order to define these conditions, three efficiencies must be

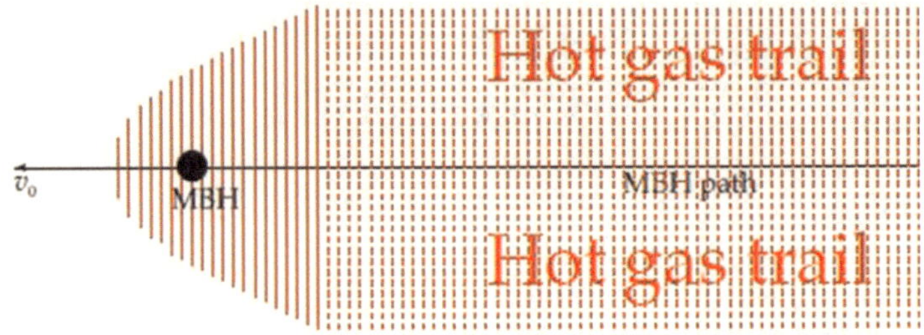

Figure 1.
MBH passage through matter.

defined. These are gas redistribution efficiency, radiative efficiency, and accretion efficiency. **Gas redistribution efficiency**, η_G, is the ratio of the accelerating force caused by gas rarefaction behind the MBH to the theoretical maximum of such force. The exact definition starts at paragraph containing Eq. (9) and ends with a paragraph containing Eq. (11). **Radiative efficiency**, η_Γ, is the ratio of the total power radiated by MBH to the power $(\dot{M}c^2)$ of the mass falling into MBH. It is expressed in Eq. (12). **Accretion efficiency**, η_A, is the ratio of the actual and the zero-radiation mass capture rates. It is defined in Eq. (16).

We show that in the case of MBH moving through stellar material at supersonic (supersonic MBH) speed, the condition for MBH acceleration is given in Eq. (29):

$$\mathcal{N} = \frac{\eta_A \eta_\Gamma \eta_{G_2}}{T_6} \gtrsim 4 \cdot 10^{-4} \left(1 + \mathcal{M}^{-2}\right)^{3/2}, \tag{1}$$

where T_6 is the temperature of the stellar material in millions Kelvin. Even though we do not have precise values for efficiencies η_A, η_Γ, and η_G, we are almost certain that for supersonic MBH, condition Eq. (1) is never met. In the case of MBH moving through stellar material at a subsonic speed (subsonic MBH), the condition for MBH acceleration is given in Eq. (33):

$$\mathcal{N} = \frac{\eta_A \eta_\Gamma \eta_{G_2}}{T_6} \gtrsim 9 \cdot 10^{-7} \mathcal{M}^3 \mathfrak{F}(\mathcal{M}, \eta_A), \tag{2}$$

where $\mathcal{M}$ is the Mach number and $\mathfrak{F}(\mathcal{M}, \eta_A) > 0.11$ is given in Eq (34). Supersonic MBH always experiences deceleration within stellar material. Subsonic MBH experiences acceleration when the Mach number exceeds $\mathcal{M}_0$ (the equilibrium Mach number) and deceleration when the Mach number is below $\mathcal{M}_0$. Eventually the MBH settles into an intrastellar orbit with Mach number $\mathcal{M}_0$. The value of $\mathcal{M}_0$ can be obtained by solving Eq. (2) as an equality.

In Appendix A, a minimal value of η_G for subsonic MBH is estimated. Estimating η_G for supersonic MBH remains an open problem. Calculating the values of η_A and η_Γ also remain open problems. As we discuss later in this work, different theorists obtained different results for η_Γ.

We briefly outline the content of the present chapter. In Section 2, we calculate forces acting on MBH. We also derive conditions for MBH acceleration at subsonic and supersonic speed. In Section 3, we present estimates for η_A and η_Γ. In Section 4, we present an empirical discussion of possible behaviors of MBH within stellar material. In Section 5, the problems remaining after this work are briefly described.

2. Forces acting on an MBH passing through matter

2.1 Total force acting on an MBH

Three forces act on a black hole, which passes through stellar material. The first force denoted by F_t is the tidal or gravitational drag. For a supersonic MBH, F_t is given by ([6], p. 8)

$$F_t = -\frac{1}{v_0}P_t = -\frac{4\pi(MG)^2\rho}{v_{0_2}}\ln\left(\frac{r_{\max}}{r_{\min}}\right), \tag{3}$$

where P_t is the decelerating power produced by the drag force, ρ is the density of the surrounding medium, $r_{\max}$ is the approximate distance from the MBH to the farthest location where the stellar material is consistent, and $r_{\min}$ is the radius at which matter is initially unperturbed by the MBH radiation. In Eq. (3), M is the mass of the MBH. We take $r_{\max}$ to be about $5 \cdot 10^7 \ m$ for a Sun-like star. We take $r_{\min}$ to be about $0.1\,m$. Hence,

$$\ln\left(\frac{r_{\max}}{r_{\min}}\right) \approx 20. \tag{4}$$

For an MBH traveling at Mach number $\mathcal{M} \lesssim 0.8$, the gravitational drag can be given by the following formula (see [7], p. 5, [8], p. 69, [9], p. 8)

$$F_t = -\frac{4\pi(MG)^2\rho}{v_{0_2}}\left[\frac{1}{2}\ln\left(\frac{1+\mathcal{M}}{1-\mathcal{M}}\right) - \mathcal{M}\right]. \tag{5}$$

Since $\mathcal{M} < 1$, Eq. (5) can be rewritten in the form of a converging series

$$F_t = -\frac{4\pi(MG)^2\rho}{v_{0_2}}\sum_{n=1}^{\infty}\frac{\mathcal{M}^{2n+1}}{2n+1}. \tag{6}$$

The second force is drag caused by mass acquisition. As the MBH passes through stellar material, it consumes mass that was formerly at rest. MBH momentum does not change as a result of mass acquisition. Change of MBH speed can be calculated from conservation of momentum:

$$\frac{\partial p}{\partial t} = \frac{\partial}{\partial t}(Mv_0) = \dot{M}v_0 + M\dot{v}_0 = 0 \qquad \Rightarrow \qquad \dot{v}_0 = -v_0\frac{\dot{M}}{M}. \tag{7}$$

Using MBH speed change, we calculate the effective force as

$$F_m = M\frac{\partial v_0}{\partial t} = -\dot{M}v_0. \tag{8}$$

The third force is accelerative. It is caused by matter rarefaction behind the moving MBH. This force is denoted by F_r. In order to estimate F_r, we need two radii, r_1 and r_2. The radius r_1 is defined in terms of F_r. A sphere of gas directly behind the MBH having radius r_1, and density $\rho/2$ would cause the MBH to experience accelerative force F_r. The sphere of rarefied gas behind an MBH acts as a sphere with a negative density of $\rho/2 - \rho = -\rho/2$. The accelerating "ramjet" force F_r acting on the MBH expressed in terms of r_1 is:

$$F_r = -\frac{MM_sG}{r_{1_2}} = MG\frac{\frac{1}{2}\frac{4}{3}\pi r_{1_3}\rho}{r_{1_2}} = \frac{2}{3}\pi MG\rho r_1, \tag{9}$$

where M_s is the effective negative mass of the sphere of rarefied gas.

The radius r_2 is defined in terms of the power P radiated by MBH passing through the stellar material. Imagine that power P is used to uniformly heat a cylinder of stellar material along the path of MBH. The MBH moves at speed v_0. The radius r_2 is defined as the radius of the aforementioned cylinder for which the temperature of gas contained in it would double. Then the relation between P and r_2 is:

$$P = (\text{Mass heated per unit of time}) \cdot T \cdot C_v = v_0\left(\pi r_{2_2}\rho\right)T \cdot C_v = \pi v_0 \rho T C_v r_{2_2}, \tag{10}$$

where C_v is the heat capacity of gas of stellar material at constant volume.

The **gas redistribution efficiency** is defined as

$$\eta_G = \frac{r_1}{r_2}. \tag{11}$$

As we show later in this section, the ramjet force F_r acting on MBH is proportional to η_G. The minimal value for η_G for subsonic MBH is estimated in Appendix A.

The radiative power of the MBH passing through stellar material is

$$P = \eta_\Gamma c^2 \dot{M}, \tag{12}$$

where η_Γ is the **radiative efficiency** of MBH and $\dot{M}$ is the mass accretion rate. For a supersonic MBH, the Bondi-Hoyle-Lyttleton accretion rate is ([10], p. 203)

$$\dot{M}_{BH} = 4\pi r_{b_2}\rho\sqrt{v_{0_2} + v_{s_2}}\ , \tag{13}$$

where r_b is the Bondi radius, and v_s is the sound speed in the stellar material. The Bondi radius is ([10], p. 203)

$$r_b = \frac{MG}{v_{s_2} + v_0}. \tag{14}$$

Substituting Eq. (14) into Eq. (13), we obtain

$$\dot{M}_{BH} = 4\pi r_b^2 \rho v_0 = \frac{4\pi (MG)^2 \rho}{\left(v_0^2 + v_s^2\right)^{3/2}}. \tag{15}$$

The actual mass capture rate is considerably smaller. The radiative heating of the gas surrounding MBH increases its temperature. This increases the gas sound speed and decreases gas density. Thus, the actual mass capture rate is

$$\dot{M} \approx \frac{4\pi (MG)^2 \rho_r}{\left(v_0^2 + v_{sr}^2\right)^{3/2}}, \tag{16}$$

where v_{sr} is the sound speed at the accretion radius and ρ_r is the density at the accretion radius. Recall the **accretion efficiency** η_A is the quotient of actual and zero-radiation mass capture rates:

$$\eta_A = \frac{\dot{M}}{\dot{M}_{BH}} \approx \left(\frac{v_{0_2} + v_{s_2}}{v_{0_2} + v_{sr_2}}\right)^{3/2} \frac{\rho_r}{\rho}. \tag{17}$$

Equating the power from Eqs. (10) and (12), we obtain

$$\pi v_0 \rho T C_v r_{2_2} = \eta_\Gamma c^2 \dot{M}. \tag{18}$$

Substituting Eqs. (15) and (16) into Eq. (18), we obtain

$$\pi v_0 \rho T C_v r_{2_2} = \eta_\Gamma c^2 \eta_A \frac{4\pi (MG)^2 \rho}{\left(v_{0_2} + v_{s_2}\right)^{3/2}}. \tag{19}$$

Thus,

$$r_2 = 2\sqrt{\eta_A}\sqrt{\frac{\eta_\Gamma c^2}{TC_v}} \frac{MG}{\left(v_{0_2} + v_{s_2}\right)^{3/4} \sqrt{v_0}} = 2\sqrt{\frac{\eta_A \eta_\Gamma c^2}{TC_v} \frac{MG}{v_{0_2}} \left(1 + \frac{v_{s_2}}{v_{0_2}}\right)^{-3/4}}. \tag{20}$$

Substituting Eq. (20) into Eq. (11), we obtain an expression for r_1:

$$r_1 = 2\eta_G \sqrt{\frac{\eta_A \eta_\Gamma c^2}{TC_v} \frac{MG}{v_{0_2}} \left(1 + \frac{v_{s_2}}{v_{0_2}}\right)^{-3/4}}. \tag{21}$$

At this point, we calculate the second and the third forces acting on the MBH. The first one is given in Eq. (3) for a supersonic MBH and in Eq. (5) for a subsonic MBH. Substituting Eq. (21) into Eq. (9), we obtain

$$F_r = \frac{2}{3}\pi MG\rho r_1 = \left[\frac{4}{3}\eta_G \sqrt{\frac{\eta_A \eta_\Gamma c^2}{TC_v} \left(1 + \frac{v_{s_2}}{v_{0_2}}\right)^{-3/4}}\right] \frac{\pi (MG)^2 \rho}{v_{0_2}}. \tag{22}$$

Substituting Eqs. (15) and (16) into Eq. (8), we obtain

$$\begin{aligned} F_m &= -\dot{M} v_0 = -\eta_A \dot{M}_{BH} v_0 = -\eta_A \frac{4\pi (MG)^2 \rho}{\left(v_{0_2} + v_{s_2}\right)^{3/2}} v_0 \\ &= -4\eta_A \left(1 + \frac{v_{s_2}}{v_{0_2}}\right)^{-3/2} \frac{\pi (MG)^2 \rho}{v_{0_2}}. \end{aligned} \tag{23}$$

2.2 Conditions for supersonic MBH acceleration

The total force acting on a supersonic MBH is obtained by summing Eqs. (3), (22) and (23):

$$F = F_t + F_m + F_r$$
$$= \frac{\pi(MG)^2\rho}{v_{0_2}}\left[-4\ln\left(\frac{r_{\max}}{r_{\min}}\right) - 4\eta_A\left(1+\frac{v_{s_2}}{v_{0_2}}\right)^{-3/2} + 2\eta_G\sqrt{\frac{\eta_A\eta_\Gamma c^2}{TC_v}}\left(1+\frac{v_{s_2}}{v_{0_2}}\right)^{-3/4}\right]. \tag{24}$$

The above equation shows that MBH accelerates if and only if $F > 0$, i.e.

$$\eta_G\sqrt{\frac{\eta_A\eta_\Gamma c^2}{TC_v}}\left(1+\frac{v_{s_2}}{v_{0_2}}\right)^{-3/4} > 2\ln\left(\frac{r_{\max}}{r_{\min}}\right) + 2\eta_A\left(1+\frac{v_{s_2}}{v_{0_2}}\right)^{-3/2}. \tag{25}$$

In this subsection we estimate conditions under which the MBH passing through matter accelerates, i.e., Eq. (25) holds. This condition can be rewritten as

$$\eta_A\eta_\Gamma\eta_{G_2} > \frac{4TC_v}{c^2}\left[\ln\left(\frac{r_{\max}}{r_{\min}}\right) + \eta_A\left(1+\frac{v_{s_2}}{v_{0_2}}\right)^{-3/2}\right]^2\left(1+\frac{v_{s_2}}{v_{0_2}}\right)^{3/2} \tag{26}$$

Recalling Eq. (4), and the fact that $\eta_A < 1$, we rewrite the estimate to Eq. (26) as

$$\eta_A\eta_\Gamma\eta_{G_2} \gtrsim 1.7\cdot 10^3\cdot\frac{TC_v}{c^2}\left(1+\frac{v_{s_2}}{v_{0_2}}\right)^{3/2} = 1.7\cdot 10^3\cdot\frac{TC_v}{c^2}\left(1+\mathcal{M}^{-2}\right)^{3/2}. \tag{27}$$

The heat capacity at the constant volume of a monatomic gas is

$$C_v = \frac{3R}{2m_a}, \tag{28}$$

where m_a is the average molar mass of the gas, and R is the gas constant ($R = 8.314\ \frac{J}{\text{mol}^0 K}$). Typical stellar material consists of monatomic gas with an average particle mass of 0.62 amu ([11], p. 378). Hence, the heat capacity at constant volume for stellar material is $C_v = 2.01\cdot 10^4\ \frac{J}{\text{kg}^0 K}$. Thus, Eq. (27) can be rewritten as

$$\mathcal{N} = \frac{\eta_A\eta_\Gamma\eta_{G_2}}{T_6} \gtrsim 4\cdot 10^{-4}\left(1+\mathcal{M}^{-2}\right)^{3/2}. \tag{29}$$

As we show in Subsection 3.2, η_A is very small if the temperature of the gas at Bondi radius is high. As we discuss in Subsection 3.3, different calculations of η_Γ in previous works yield different results, yet all of them are below 0.1. For supersonic MBH, η_G should rapidly decrease with increasing Mach number. We have only qualitative arguments regarding the values of η_G. Stellar matter behind the MBH, which is displaced by a heat wave, remains within the Mach cone. Its gravitational pull can not be much lower than the pull of the unaffected matter in front of MBH. As the MBH Mach number increases, the cone becomes narrower. The difference of gravitational pull between matter in front of MBH and behind MBH decreases. Hence, η_G decreases as well. Calculation of η_G is beyond the scope of this work. The solar gas temperature exceeds $T_6 = 4$ for radius under 0.5 Solar radii [11].

Based on the above data, we can be almost certain that relation Eq. (29) does not hold for Mach numbers $\mathcal{M} > 1$, thus a supersonic MBH can not accelerate. **Very**

extensive analysis is needed in order to rigorously prove this assertion. Such analysis is beyond the scope of this work. It may be beyond the scope of any previous work on black hole accretion.

2.3 Conditions for subsonic MBH acceleration

The tidal decelerating force acting on an MBH traveling through stellar material at Mach number $\mathcal{M}\lesssim 0.8$ is given by Eq. (6). The total force acting on MBH is obtained by summing Eqs. (6), (22), and (23):

$$\begin{aligned} F &= F_t + F_m + F_r \\ &= \frac{\pi(MG)^2\rho}{v_{0_2}}\left[-4\sum_{n=1}^{\infty}\frac{\mathcal{M}^{2n+1}}{2n+1} - 4\eta_A\left(1+\frac{v_{s_2}}{v_{0_2}}\right)^{-3/2} + 2\eta_G\sqrt{\frac{\eta_A\eta_\Gamma c^2}{TC_v}}\left(1+\frac{v_{s_2}}{v_{0_2}}\right)^{-3/4}\right] \\ &= \frac{\pi(MG)^2\rho}{v_{0_2}}\left[-4\sum_{n=1}^{\infty}\frac{\mathcal{M}^{2n+1}}{2n+1} - 4\eta_A\left(1+\frac{1}{\mathcal{M}^2}\right)^{-3/2} + 2\eta_G\sqrt{\frac{\eta_A\eta_\Gamma c^2}{TC_v}}\left(1+\frac{1}{\mathcal{M}^2}\right)^{-3/4}\right]. \end{aligned} \tag{30}$$

The above equation shows that MBH will accelerate if and only if $F > 0$ or

$$\eta_G\sqrt{\frac{\eta_A\eta_\Gamma c^2}{TC_v}}\left(1+\mathcal{M}^{-2}\right)^{-3/4} > 2\sum_{n=1}^{\infty}\frac{\mathcal{M}^{2n+1}}{2n+1} + 2\eta_A\left(1+\mathcal{M}^{-2}\right)^{-3/2}. \tag{31}$$

Rewrite Eq. (31) as:

$$\begin{aligned} \eta_A\eta_\Gamma\eta_{G2} &> \frac{4TC_v}{c^2}\left(1+\mathcal{M}^{-2}\right)^{3/2}\left[\sum_{n=1}^{\infty}\frac{\mathcal{M}^{2n+1}}{2n+1} + \eta_A\left(1+\mathcal{M}^{-2}\right)^{-3/2}\right]^2 \\ &= \frac{4TC_v}{c^2}\mathcal{M}^3\left\{\left(1+\mathcal{M}^2\right)^{3/2}\left[\sum_{n=0}^{\infty}\frac{\mathcal{M}^{2n}}{2n+3} + \eta_A\left(1+\mathcal{M}^2\right)^{-3/2}\right]^2\right\}. \end{aligned} \tag{32}$$

Given that $C_v = 2.01\cdot 10^4 \ \frac{J}{kg^oK}$, we rewrite Eq. (32) as

$$\frac{\eta_A\eta_\Gamma\eta_{G_2}}{T_6} \gtrsim 9\cdot 10^{-7}\mathcal{M}^3\mathfrak{F}(\mathcal{M},\eta_A), \tag{33}$$

where

$$\mathfrak{F}(\mathcal{M},\eta_A) = \left(1+\mathcal{M}^2\right)^{3/2}\left[\sum_{n=0}^{\infty}\frac{\mathcal{M}^{2n}}{2n+3} + \eta_A\left(1+\mathcal{M}^2\right)^{-3/2}\right]^2. \tag{34}$$

Notice that $\mathfrak{F}(\mathcal{M},\eta_A) > .11$.

The Mach number for which an MBH settles into a stable intrastellar orbit is such that the net force acting on the MBH is 0. It can be estimated by solving an equation derived from Eq. (33):

$$\mathcal{N} = \frac{\eta_A \eta_\Gamma \eta_{G_2}}{T_6} = 9 \cdot 10^{-7} \mathcal{M}^3 \mathfrak{F}(\mathcal{M}, \eta_A). \tag{35}$$

All three efficiencies in Eq. (35) are nonzero. Thus, Eq. (35) does have a solution $\mathcal{M}_0$. An MBH traveling in stellar material accelerates when its Mach number is below $\mathcal{M}_0$ and decelerates when its Mach number is above $\mathcal{M}_0$. Thus, an MBH traveling within a star is bound to settle into a stable intrastellar orbit. In order to calculate $\mathcal{M}_0$ from Eq. (35), one must know gas redistribution, accretion and radiative efficiencies. In the next section, we present preliminary estimates for the three aforementioned efficiencies.

3. Estimation of gas redistribution, accretion and radiative efficiencies

3.1 The value of η_G

In Appendix A.1, we prove that

$$\eta_G = \frac{r_1}{r_2} \geq .44\sqrt{\frac{K-1}{K}} \tag{36}$$

for MBH traveling at subsonic speeds with $\mathcal{M} < 0.8$. K is the ratio of average temperature in the hot gas trail and ambient temperature of stellar material. A possible route for estimating K is outlined in Appendix A.2.

3.2 The value of η_A

From Eq. (17), we estimate η_A as

$$\eta_A \approx \left(\frac{v_{0_2} + v_{s_2}}{v_{0_2} + v_{sr_2}}\right)^{3/2} \frac{\rho_r}{\rho}, \tag{37}$$

where v_{sr} is the sound speed at the accretion radius and ρ_r is the density at the accretion radius. Given that gas density is proportional to its pressure divided by temperature, we obtain

$$\eta_A \approx \left(\frac{v_{0_2} + v_{s_2}}{v_{0_2} + v_{sr_2}}\right)^{3/2} \frac{\rho_r}{\rho} = \left(\frac{v_{0_2} + v_{s_2}}{v_{0_2} + v_{sr_2}}\right)^{3/2} \frac{\mathbf{P}_r}{\mathbf{P}} \frac{T}{T_r} = \left(\frac{(v_0/v_s)^2 + 1}{(v_0/v_s)^2 + (v_{sr}/v_s)^2}\right)^{3/2} \frac{\mathbf{P}_r}{\mathbf{P}} \frac{T}{T_r}$$
$$= \left(\frac{\mathcal{M}^2 + 1}{\mathcal{M}^2 + (v_{sr}/v_s)^2}\right)^{3/2} \frac{\mathbf{P}_r}{\mathbf{P}} \frac{T}{T_r}. \tag{38}$$

In Eq. (38), T is the ambient temperature of stellar material, and T_r is the temperature of stellar material at Bondi radius. The ambient pressure is $\mathbf{P}$, and pressure at Bondi radius is $\mathbf{P}_r$. Notation P can not be used for pressure, as it is already used for power. For a gaseous medium, the sound velocity is proportional to the square root of temperature. Thus,

$$\left(\frac{v_{sr}}{v_s}\right)^2 = \frac{T_r}{T}. \tag{39}$$

Substituting Eq. (39) into Eq. (38), we obtain the approximation

$$\eta_A \approx \left(\frac{\mathcal{M}^2 + 1}{\mathcal{M}^2 + T_r/T}\right)^{3/2} \frac{\mathbf{P}_r}{\mathbf{P}} \frac{T}{T_r}. \tag{40}$$

The pressure within the immediate vicinity of MBH should be approximated by the sum of gas pressure and dynamic pressure:

$$\mathbf{P}_r = \mathbf{P} + \frac{\rho v_{0_2}}{2} = \mathbf{P} + \rho v_{s_2} \frac{\mathcal{M}^2}{2}. \tag{41}$$

Notice that the gas density at the Bondi radius must be equal to or lower than the density of unperturbed gas. Hence, the approximation in Eq. (41) above works only when

$$\frac{\mathbf{P}_r}{\mathbf{P}} \frac{T}{T_r} < 1, \tag{42}$$

The relation between pressure and sound velocity in a monatomic ideal gas is ([12], p. 683):

$$\mathbf{P} = \frac{\rho v_{s_2}}{\gamma} = \frac{3}{5} \rho v_{s_2}. \tag{43}$$

Substituting Eq. (43) into Eq. (41), we obtain the pressure ratio

$$\frac{\mathbf{P}_r}{\mathbf{P}} = \frac{\frac{3}{5} \rho v_{s_2} + \rho v_{s_2} \frac{\mathcal{M}^2}{2}}{\frac{3}{5} \rho v_{s_2}} = 1 + \frac{5}{6}\mathcal{M}^2. \tag{44}$$

Substituting Eqs. (44) and (42) into Eq. (40), we obtain

$$\eta_A \approx \left(\frac{\mathcal{M}^2 + 1}{\mathcal{M}^2 + T_r/T}\right)^{3/2} \min\left\{1, \left(1 + \frac{5}{6}\mathcal{M}^2\right)\frac{T}{T_r}\right\}. \tag{45}$$

As we see, η_A is a rapidly increasing function of the Mach number and a rapidly decreasing function of T_r/T. For subsonic MBH and for all cases where $T_r/T \gg \mathcal{M}^2$, Eq. (45) can be approximated as

$$\eta_A \approx \left(1 + \frac{5}{6}\mathcal{M}^2\right)\left(\mathcal{M}^2 + 1\right)^{3/2}\left(\frac{T}{T_r}\right)^{5/2}. \tag{46}$$

Calculation of T_r remains an unsolved problem.

3.3 The value of η_Γ

Some energy is radiated from a spherically accreting MBH in the form of photons. The power radiated as photons is given as $\eta_\gamma \dot{M} c^2$. Some energy is radiated from a spherically accreting MBH in the form of protons and neutrons. The power radiated as baryons is given as $\eta_p \dot{M} c^2$, since protons are more numerous than neutrons. The overall radiative efficiency of an MBH is

$$\eta_\Gamma = \eta_\gamma + \eta_p. \tag{47}$$

3.3.1 Gamma radiation from spherically accreting MBH

Accretion rate per unit MBH mass is one of the main factors determining η_γ. This rate should be expressed as a multiple of the Eddington accretion rate ([13], p. 51):

$$\mathfrak{A} = \frac{\left(1.43 \cdot 10^{16}\ s\right)\ \dot{M}}{M} = \frac{\dot{M}}{\left(70\ \frac{kg}{s}\right) M_{18}}, \tag{48}$$

where M_{18} is the mass of MBH in units of 10^{18} kg.

Below we will summarize some previous works calculating η_γ for spherical accretion. Spherical accretion on black holes have been studied theoretically, with different theories producing different values of radiative efficiency (η_γ) ([13], p. 25–55). Radiative efficiencies ranging from 10^{-10} to over .1 have been obtained for different parameters. The magnetic field greatly increases η_γ ([13], p. 34–35). For $10^{-4} \le \mathfrak{A} \le 1$, radiative efficiency can be as high as 0.1 if the flow is turbulent ([13], p. 35).

Detailed calculations of spherical accretion are presented in Ref. [14]. For a black hole of $2 \cdot 10^{38}$ kg, radiative efficiency starts growing almost from zero at $\mathfrak{A} = .02$ and reaches $\eta_\gamma = .19$ for $\mathfrak{A} = 1.2$. For a black hole of $2 \cdot 10^{31}$ kg, radiative efficiency starts growing almost from zero at $\mathfrak{A} = .5$ and reaches $\eta_\gamma = .15$ for $\mathfrak{A} = .12$. MBH was not considered.

A model which considers separate ion and electron temperatures within accreting gas is given in Ref. [15]. Black hole masses between2 $\times$ 10^{31} kg and 2×10^{38} kg are considered. Accretion rates between $\mathfrak{A} = 7 \cdot 10^{-3}$ and $\mathfrak{A} = 2$ are considered. In all cases, the efficiency stays within $\eta_\gamma \in \left[4.8 \cdot 10^{-3}, 7 \cdot 10^{-3}\right]$. Notice, that all of the aforementioned studies considered black holes many orders of magnitude heavier than 10^{18} kg. To obtain better results for MBH, more detailed studies for black holes within 10^{16} kg–3×10^{19} kg are needed.

For black holes with accretion rates $\mathfrak{A} \in (1, 300)$, the values of η_γ range from 10^{-6} to 10^{-2} ([16], p.10). The state-of-the-art results have a lot of uncertainty.

3.3.2 Proton and neutron radiation from spherically accreting MBH

Gas accreting toward MBH experiences great compression, which causes adiabatic heating. Hot gas reaches temperatures of tens to hundreds of billion degrees Kelvin. As a result, some protons and neutrons which have excess energy escape the gravitational well around MBH. A very rudimentary estimation of η_p is performed below. In order to calculate η_p precisely, we would need to perform an extensive Monte Carlo simulation. This simulation would have to take into account proton motion and collisions.

During accretion, the electron gas is much colder than the proton gas. Average temperature of proton is approximated by ([14], p. 17, [15], p. 323):

$$T(yr_s) = \frac{T_s}{y}, \tag{49}$$

where r_s is the Schwarzschild radius and $T_s \approx 10^{12o}K$. When the distance from MBH is corresponding to $y \in (25, 50)$ Schwarzschild radii and the gas temperature is 20–40 billion Kelvin, the nuclei split into protons and neutrons.

At this point, we calculate the depth of the potential well in which nucleons appear at a distance (yr_s) from the MBH center. We take the non-relativistic approximation valid for $y \geq 2$.

$$E_p(yr_s) = -\frac{m_p\ MG}{yr_s} = -\frac{m_p\ MG}{y\frac{2MG}{c^2}} = -\frac{m_p\ c^2}{2y}, \tag{50}$$

where m_p is the proton mass. Below, we express Eq. (50) in terms of Boltzmann constant $k = 1.381 \times 10^{-23}$ J/K:

$$E_p(yr_s) = -\frac{m_p\ c^2}{2y} = -\frac{k}{y}\cdot\frac{m_p\ c^2}{2k} = -5.44\cdot 10^{12o}K\ \frac{k}{y} \approx -\frac{5.44\ k\ T_s}{y}. \tag{51}$$

Like particles of any gas, protons and neutrons within accreting gas should have Maxwell energy distribution:

$$f(E) = \frac{2}{\sqrt{\pi}\ kT}\sqrt{\frac{E}{kT}}\ \exp\left(-\frac{E}{kT}\right), \tag{52}$$

At any distance (yr_s) from the MBH center, some nucleons have sufficient kinetic energy to escape from the gravitational potential well of MBH. The energy depth of that well is given by Eq. (51). The fraction of nucleons capable of escaping is

$$\mathcal{F}_e = \int_{5.44}^{\infty} f(E)dE = 0.012. \tag{53}$$

Nucleons escaping from a distance (yr_s) from the MBH center carry excess kinetic energy. That energy is

$$\mathcal{F}_e = \frac{kT_s}{y}\int_{5.44}^{\infty}(E - 5.44)f(E)dE = 0.009\ \frac{kT_s}{y}. \tag{54}$$

The energy given in Eq. (54) above is the quotient of the excess energy of ejected nucleons to the total number of nucleons, including the ones not ejected.

Define η_p^* as the quotient of the kinetic energy of nucleons ejected from accreting material to the rest energy of all nucleons. Many ejected nucleons lose energy in collisions, and some return to MBH. Thus, the final energy radiated from MBH as nucleon radiation is $\eta_p < \eta_p^*$. We estimate η_p^* as

$$\frac{d\eta_p^*}{d(lny)} = \frac{\mathcal{F}_e(y)}{m_p c^2} = \frac{0.009\ kT_s}{m_p c^2}\ \frac{1}{y} \approx \frac{8\cdot 10^{-4}}{y}. \tag{55}$$

Integrating Eq. (55) for $y > 2$, we obtain

$$\eta_p^* \approx \int_{y=2}^{\infty} \frac{8 \cdot 10^{-4}\, dy}{y} = 4 \cdot 10^{-4}. \tag{56}$$

The value of η_p depends on the fraction of the nucleons which are slowed down by accreting gas and returning to MBH. An extensive study and simulation may yield the value of η_p higher than the value of η_p^* estimated in Eq. (56). At this point, precise efficiencies are unknown.

4. Possible modes of interaction of MBH with a star

In this section, we discuss the behavior of a Primordial Black Hole (PBH) which is captured into an orbit that intersects a star. Every PBH discussed here is an MBH, since it is microscopic. Not every MBH is a PBH, since some MBH are not primordial.

PBH ejection from a star-intersecting orbit is the first mode of PBH-star interaction. Any MBH or PBH on a star-intersecting orbit moves within stellar material with supersonic speed. Thus, it experiences deceleration within stellar material. Such MBH or PBH can be ejected from its orbit only by gravitational interaction with the star's planets. In our opinion, such ejections are not rare. Reasoning follows.

Kinetic energy loss of an MBH on a single intrastellar passage is (see Appendix B):

$$\Delta E_{\text{pass}} = \left(2.0 \cdot 10^{19}\ J\right) M_{18^2}. \tag{57}$$

The energy needed to drop the apogee of an elliptic MBH orbit around a sun-like star to 1 Astronomical Unit is

$$\Delta E_{\text{orbit}} = \frac{G M_{\text{Sun}} M_{\text{MBH}}}{1\ AU} = \left(8.9 \cdot 10^{26}\ J\right) M_{18}. \tag{58}$$

Dividing Eq. (58) by Eq. (57), we obtain the number of times an MBH has to pass through a star in order for its orbit apogee to descend to 1 AU:

$$N = \frac{\Delta E_{\text{orbit}}}{\Delta E_{\text{pass}}} \approx 4.5 \cdot 10^{7} (M_{18})^{-1}. \tag{59}$$

During this number of passes, the gravity of satellites of a star may throw an MBH off the orbit.

Settling of MBH into an intrastellar orbit is the second mode of MBH-star interaction. One possibility of MBH entering an intrastellar orbit is an MBH is a capture by a star. Another possibility is MBH production at the star center by coalescence of dark matter [4, 5]. Such MBH would be accelerated until it settles in an intrastellar orbit.

Consumption of a host star by an MBH is the third mode of MBH-star interaction. The evolution of an intrastellar MBH depends on its growth rate. An intrastellar MBH moves at low subsonic speed, hence its mass growth rate can be approximated by Eq. (15) which holds for a stationary MBH:

$$\dot{M} = \frac{4\pi (MG)^2 \rho_r}{v_{sr_3}}, \tag{60}$$

where ρ_r is the density at Bondi radius and v_{sr} is the sound speed and Bondi radius. Sound velocity within the gas is proportional to $T^{0.5}$. Gas density is proportional to T^{-1}. Hence,

$$\dot{M} = \frac{4\pi(MG)^2\rho}{v_{s3}}\left(\frac{T}{T_r}\right)^{2.5}, \tag{61}$$

where ρ, v_s, and T are the density, sound speed, and temperature of stellar material, while T_r is the temperature at Bondi radius. In the Solar center, the density is $1.5\times10^5\ \frac{\mathrm{kg}}{\mathrm{m}^3}$ and the sound speed is $5.1\times10^5\ \frac{\mathrm{m}}{\mathrm{s}}$ ([11], p. 378). Substituting the above into Eq. (61), we obtain the accretion rate

$$\dot{M} = 6.3\times10^4\ \frac{\mathrm{kg}}{\mathrm{s}}\left(\frac{T}{T_r}\right)^{2.5} M_{18}^2 = \frac{2\times10^{18}\ \mathrm{kg}}{\text{Million years}}\left(\frac{T}{T_r}\right)^{2.5} M_{18}^2. \tag{62}$$

Dividing both sides of the above equation by mass, we obtain

$$\frac{d}{dt}(\ln M_{18}) = \frac{\dot{M}_{18}}{M_{18}} = \frac{\dot{M}}{M} = \frac{2M_{18}}{\text{Million years}}\left(\frac{T}{T_r}\right)^{2.5}. \tag{63}$$

In Eq. (63) above, T and T_r are gas temperatures of ambient matter and at Bondi radius respectively. A low mass MBH is unlikely to experience significant growth over the lifetime of the host star. Determining exact MBH and host star characteristics for which the host star is consumed remains an open problem.

As we see from Eq. (63), the initial growth of an MBH within a star is slow. As the MBH gains mass with $M_{18} > 100$, all emitted radiation is absorbed by the accreting gas and $T \approx T_r$. Then the star is consumed by MBH over several millennia.

The growth of intrastellar black holes has been considered by previous researchers [17]. As a black hole consumes a star, it obtains the star's angular momentum and becomes a rapidly rotating black hole. As a rotating black hole absorbs matter, it radiates two jets along its axis [18]. The final stages of stellar consumption by MBH may be responsible for long γ-ray pulses [19, 20].

5. Conclusion and remaining problems

In this work, we have demonstrated that MBH passing through stellar material experiences acceleration rather than deceleration as long as

$$\mathcal{N} = \frac{\eta_A\eta_\Gamma\eta_{G_2}}{T_6} \gtrsim \left\{\begin{array}{ll} 4\cdot10^{-4}\ \left(1+\mathcal{M}^{-2}\right)^{3/2} & \text{for supersonic MBH} \\ 9\cdot10^{-7}\mathcal{M}^3\mathfrak{F}(\mathcal{M},\eta_A) & \text{for subsonic MBH}\end{array}\right|, \tag{64}$$

where $\mathfrak{F}(\mathcal{M},\eta_A)$ is given in Eq. (34). T_6 is the temperature of the stellar material in millions Kelvin. The **gas redistribution efficiency** η_G, **radiative efficiency** η_Γ, and **accretion efficiency** η_A are defined in Introduction and Section 2.

MBH in stellar material experiences deceleration at supersonic speed. Subsonic MBH either accelerates or decelerates until it reaches equilibrium Mach number calculated from (??) and settles into a stable intrastellar orbit.

If the Universe contains MBH, many or most of them may exist in intrastellar orbits within stars. Some MBH may be orbiting within the Sun. Some of these MBH may be PBH captured by stars. We do not know how frequent is stellar capture of PBH. The calculation of this frequency is one of the many open problems generated by this work. Different PBH masses as well as star and planetary system characteristics will have to be considered in this calculation.

Other MBH may be generated within stellar centers. According to some theories, most Dark Matter consists of Weakly Interacting Massive Particles (WIMPs). Within stellar centers, WIMPs may coalesce into MBH [4, 5]. These MBH would experience acceleration until they settle into intrastellar orbits.

Several detectable effects may be produced by MBH on intrastellar orbits. Some Type 1a supernovas may be triggered by these MBHs [4]. Some MBHs may be on an intrastellar orbit within Sun. These MBH produce very low-frequency sonic waves. These waves are detectable by **helioseismology**—study of vibrations of Solar photosphere.

Only very low frequency sound can travel long distances in any gas. Sound with a frequency of a few millihertz or lower can travel from the Solar center to the Solar surface [21]. From the data presented in Ref. ([11], p. 378) we calculate that the orbital period of an MBH on an intasolar orbit is at least 800 s. This shows that acoustic waves produced by MBH rich Solar surface. Hence, these waves can be detected.

As mentioned in Subsection 3.3, **radiative efficiencies** η_Γ of accreting MBH can not be determined at this point. Most advanced theories give results, which vary by several orders of magnitude. Values ranging from 10^{-10} to 0.1 have been obtained so far. We do not know which theory is correct. If one or more MBH orbiting within the Sun is detected, then true values of radiative efficiencies will be obtained from observation.

In Appendix A, we estimate a minimum value of η_G for subsonic MBH. The exact calculation of η_G is a remaining problem. It would involve extensive theoretical work and simulations using gas dynamics and radiation-matter interaction.

Accretion efficiency η_A for both subsonic and supersonic MBH is given by Eq. (45) in terms of T_r—the temperature at the Bondi radius. The calculation of T_r is a remaining problem. Exact calculation of T_r, and η_A would involve extensive theoretical work and simulations using gas dynamics, radiation energy transport, and magnetohydrodynamics.

This work is purely theoretical. Nevertheless, helioseismological observations may eventually provide evidence of an MBH orbiting in an itrasolar orbit. This observation may open possibilities to obtain additional knowledge in many branches of physics. Knowledge in any branch of physics may lead to unforeseeable technological advances in the future.

A. Estimation of bounds on η_G

A.1 Minimum value of η_G for subsonic MBH

We assume strictly subsonic regime with $\mathcal{M} \leq 0.8$. In a diagram below, we illustrate the MBH passing through stellar material.

The heated stellar material produced by subsonic MBH consists of two regions. The first region is the parabolic head region of hot gas surrounding the MBH. The second region is the hot gas trail, denoted by $\mathcal{R}_{hg}$.

An important issue is the location of stellar material mass displaced by the heat wave. If MBH has subsonic speed, then sonic density waves carry away all of the displaced mass. Sonic waves are shaped as expanding spherical shells. Each shell is centered at the point of wave origin. A subsonic MBH can not outrun shells expanding at the speed of sound. Hence, all of the expanding shells contain advancing MBH inside them. As a result, these shells containing displaced matter exert no net gravitational force on MBH.

Accelerative force F_r exerted on MBH comes from the difference in density of ambient stellar material and hot rarefied gas within the tail region $\mathcal{R}_{hg}$ as well as the head region as shown on **Figure 2**. We simplify calculation by ignoring the head region, which does provide a small propulsive force.

The gravitational force exerted by $\mathcal{R}_{hg}$ is calculated below. The region $\mathcal{R}_{hg}$ can be approximated by a cylinder with radius r_h. This cylinder starts at the distance at most r_h from the MBH. By taking the distance to be r_h, we are estimating the minimal value of the force. The region $\mathcal{R}_{hg}$ is represented in cylindrical coordinates with MBH at the origin. The direction in which the MBH is traveling is $-\hat{z}$. In cylindrical coordinates, $\mathcal{R}_{hg}$ is given by

$$\begin{cases} z \in (r_h, \infty) \\ r \in [0, r_h) \end{cases} \tag{A.1}$$

Stellar gas temperature in $\mathcal{R}_{hg}$ is approximated by a uniform temperature KT, where $K > 1$ is a constant and T is the ambient temperature. Gas pressure within $\mathcal{R}_{hg}$ is almost the same as ambient gas pressure. From temperature and pressure in $\mathcal{R}_{hg}$, it follows that the gas density in that region is ρ/K, where ρ is the ambient density. From the standpoint of gravitational interaction, **effective negative density** ρ_- of the material in $\mathcal{R}_{hg}$ can be defined as the difference between gas density in $\mathcal{R}_{hg}$ and ambient gas density. Effective negative density is

$$\rho_- = \frac{\rho}{K} - \rho = -\rho\ \frac{K-1}{K}. \tag{A.2}$$

The rarefied gas region $\mathcal{R}_{hg}$ exerts the following force on MBH:

$$\mathbf{F} = \iiint_{\mathcal{R}_{hg}} (\rho_-)\mathbf{g}(\mathbf{r})dV = -\rho\ \frac{K-1}{K} \iiint_{\mathcal{R}_{hg}} \mathbf{g}(\mathbf{r})dV. \tag{A.3}$$

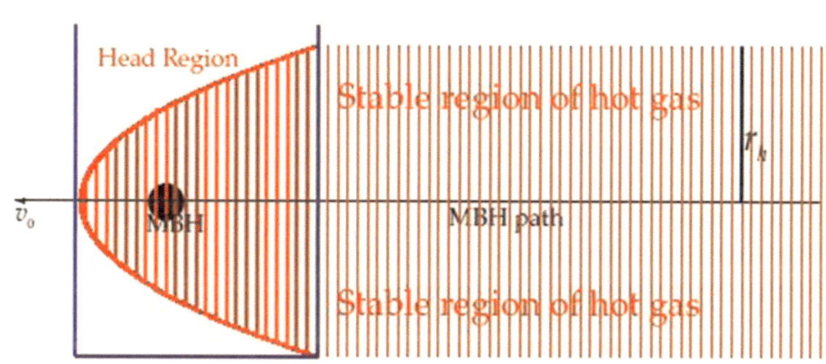

Figure 2.
Heat wave caused by subsonic MBH.

In Eq. (A.3) above, acceleration due to MBH gravity at point **r** is

$$\mathbf{g}(\mathbf{r}) = -MG\frac{\mathbf{r}}{\|\mathbf{r}\|^3} \tag{A.4}$$

Substituting Eq. (A.4) into Eq. (A.3), we obtain

$$\mathbf{F} = -MG\rho\ \frac{K-1}{K}\iiint_{\mathcal{R}_{hg}}\frac{\mathbf{r}}{\|\mathbf{r}\|^3}dV, \tag{A.5}$$

The gas displaced by the MBH passage in $-\hat{z}$ direction retains cylindrical symmetry. This symmetry implies that the net force on the MBH will act only in $-\hat{z}$ direction. Thus, Eq. (A.5) can be further simplified to

$$\begin{aligned}
F_r &= -\mathbf{F}\cdot\hat{z} = MG\rho\ \frac{K-1}{K}\iiint_{\mathcal{R}_{hg}}\frac{\mathbf{r}\cdot\hat{z}}{\|\mathbf{r}\|^3}dV = MG\rho\ \frac{K-1}{K}\iiint_{\mathcal{R}_{hg}}\frac{z}{(z^2+r^2)^{3/2}}\ dV \\
&\geq MG\rho\ \frac{K-1}{K}\int_0^{r_h}\int_{r_h}^{\infty}\frac{\pi rz}{(z^2+r^2)^{3/2}}dzdr = \pi MG\rho\ \frac{K-1}{K}\int_0^{r_h}\left[-\frac{r}{\sqrt{z^2+r^2}}\right]_{z=r_h}^{z=\infty}dr \\
&= \pi MG\rho\ \frac{K-1}{K}\int_0^{r_h}\frac{r}{\sqrt{r_{h_2}+r^2}}dr = \pi MG\rho\ \frac{K-1}{K}\left[\sqrt{r_{h_2}+r^2}\right]_{r=0}^{r_h} \\
&= \pi MG\rho\ r_h\left(\frac{K-1}{K}\left(\sqrt{2}-1\right)\right).
\end{aligned} \tag{A.6}$$

As we have mentioned earlier, the real force is greater or equal to the one calculated by approximating $\mathcal{R}_{hg}$ by Eq. (A.1). Eqs. (A.6) and (9), we obtain

$$r_1 \geq \frac{3}{2}\left(\sqrt{2}-1\right)\ \frac{K-1}{K}\ r_h \approx 0.62\ \frac{K-1}{K}\ r_h. \tag{A.7}$$

Below, r_h is estimated in terms of r_2. The power needed to heat the gas trail is

$$\begin{aligned}
P_T &= (\text{Mass heated per unit of time})\cdot(\text{Temperature})\cdot C_p \\
&= v_0\left(\pi r_{h_2}\frac{\rho}{K}\right)((K-1)T)\left(\frac{5}{3}C_v\right) = \frac{5(K-1)}{3K}\pi v_0\rho TC_v r_{h_2},
\end{aligned} \tag{A.8}$$

where P_T is the thermal power. For monatomic gas, $C_p = \frac{5}{3}C_v$. Some of the power P radiated by the MBH goes into the production of the sonic waves, hence $P > P_T$. Accurate calculation of P_T/P is beyond the scope of this work. Nevertheless, for subsonic MBH, we are certain that no more than 20% of MBH heating power is consumed by making sonic waves. Therefore, $P_T/P \gtrsim 0.8$. Using this data, we estimate the total radiative power of MBH:

$$P \leq \frac{2(K-1)}{K}\pi v_0\rho TC_v r_{h_2} \tag{A.9}$$

Substituting Eq. (A.9) into Eq. (10), we obtain

$$\frac{2(K-1)}{K}\pi v_0 \rho T C_v r_{h_2} \geq \pi v_0 \rho T C_v r_{2_2}. \tag{A.10}$$

Hence,

$$r_2 \leq r_h \sqrt{\frac{2(K-1)}{K}}. \tag{A.11}$$

Dividing Eq. (A.7) by Eq. (A.11), we obtain

$$\eta_G = \frac{r_1}{r_2} \geq .44 \sqrt{\frac{K-1}{K}}. \tag{A.12}$$

Calculation of K is beyond the scope of this work. Some considerations regarding the value of K are presented in Appendix A.2.

A.2 Estimation of K

Recall, that the average temperature of the gas in the hot tail is KT, where T is the temperature of the surrounding stellar material. In order to make any inference on the value of K, we introduce two radii and calculate their ratio. **Radiation radius** r_γ is the average distance traveled by a photon or another energy-carrying particle from PBH before being absorbed by stellar material. **Minimal hot tail radius** r_{mh} is the minimal radius the hot tail can have regardless of K.

Below we estimate r_γ and r_{mh}. The radiation radius is

$$r_\gamma = \frac{\mathfrak{S}_\gamma}{\rho_p} = \frac{\mathfrak{S}_\gamma}{10^3 \ \frac{kg}{m^3} \ \rho_{3p}}, \tag{A.13}$$

where $\mathfrak{S}_\gamma$ is the planar density of material through which an energy carrying particle has to travel before being absorbed by stellar material. The density ρ_p is an average density of the material over the path of the energy-carrying particle, and ρ_{3p} is the same density in $10^3 \mathrm{kg/m^3}$. The value of $\mathfrak{S}_\gamma$ is inversely proportional to average absorption cross-section of the energy-carrying particles:

$$\mathfrak{S}_1 = \frac{1 \text{ kg}}{1000 \ N_A \text{ amu}} \cdot \frac{1}{10^{-28} \ \sigma} = \frac{17 \ \frac{\text{kg}}{\text{m}^2}}{\sigma \text{ in barn}}, \tag{A.14}$$

where σ is the absorption cross-section. Given that most interactions are scattering, effective absorption cross-section has to be calculated. Substituting Eq. (A.14) into Eq. (A.13), we obtain

$$r_\gamma = \frac{\mathfrak{S}_\gamma}{\rho_p} = \frac{0.017 \text{ m}}{\rho_{3p}(\sigma \text{ inbarn})}. \tag{A.15}$$

According to data presented in ([22], pp. 41–42), cross-section per amu decreases with photon energy. For 10 keV photon, it is 0.55 barn For 1 MV photon, it is 0.18 barn. For 50 MV photon, it is 0.023 barn.

The minimal hot tail radius can be obtained from Eq. (A.8):

$$r_{mh} = \sqrt{\frac{3P_T}{5\pi v_0 \rho T C_v}}, \tag{A.16}$$

where P_T is the part of MBH power used to produce heat rather than the sound wave. Substituting Eqs. (15)–(17) into Eq. (12) we obtain

$$P_T = \eta_h P = \eta_h \eta_\Gamma \eta_A \frac{4\pi (MG)^2 c^2 \rho}{\left(v_{0_2} + v_{s_2}\right)^{3/2}} = \eta_h \eta_\Gamma \eta_A \frac{4\pi (MG)^2 c^2 \rho}{v_{0_3}\left(1 + \mathcal{M}^{-2}\right)^{3/2}}, \tag{A.17}$$

where $\eta_h \approx 0.8$ is the fraction of MBH radiative power which goes into heating the stellar medium rather than producing a sonic wave. Substituting Eq. (A.17) into Eq. (A.16), we obtain

$$r_{mh} \approx 1.5\ \sqrt{\eta_\Gamma \eta_A} \frac{MGc\left(1 + \mathcal{M}^{-2}\right)^{-3/4}}{v_{0_2}\sqrt{TC_v}} \tag{A.18}$$

As mentioned in Subsection 2.2, for average stellar material, $C_v = 2.01 \cdot 10^4\ \frac{J}{kg^o K}$. From Eq. (A.18), we obtain

$$r_{mh} \approx (0.21\ m)\ \sqrt{\eta_\Gamma \eta_A}\ \left(1 + \mathcal{M}^{-2}\right)^{-3/4} M_{18}\ v_{6_{-2}}\ T_{6_{-1/2}}. \tag{A.19}$$

Combining Eq. (A.15) and Eq. (A.19), we obtain the ratio

$$\mathcal{R}_\gamma = \frac{r_\gamma}{r_{mh}} \approx 0.8 \frac{v_{6_2}\ \sqrt{T_6}\ \left(1 + \mathcal{M}^{-2}\right)^{3/4}}{\rho_{3p} M_{18}(\sigma \text{ in barn})\sqrt{\eta_\Gamma \eta_A}} \approx \frac{\sqrt{v_6\ T_6}\ \left(\max\{v_6, v_{6s}\}\right)^{3/2}}{\rho_{3p} M_{18}(\sigma \text{ in barn})\sqrt{\eta_\Gamma \eta_A}}, \tag{A.20}$$

where v_{6s} is the sound velocity in 10^6 m/s.

If $\mathcal{R}_\gamma \ll 1$, then gas close to MBH is heated to a great temperature. This gas expands before it has time to diffuse its heat. The expanded gas must remain hot in order to balance the outside pressure. In that case, $K \gg 1$. For $\mathcal{R}_\gamma \gg 1$, thermal energy is dissipated over a very large gas volume. This gas volume is heated only by a small margin, thus $0 < K - 1 \ll 1$.

B. Estimation of a MBH kinetic energy loss on passage through a sun-like star

Using $r_{\min} = 0.1\ m$ and $r_{\max} = 5 \cdot 10^7\ m$ to express Eq. (3) in numerical terms we obtain:

$$F_t = \frac{4\pi (MG)^2 \rho}{v_{0_2}} \ln\left(\frac{r_{\max}}{r_{\min}}\right) = \left(1.12 \cdot 10^9\ N\right) \frac{M_{18_2}\ \rho_3}{v_{6_2}}, \tag{B.1}$$

where ρ_3 is density in 10^3 kg/m^3, and v_6 is velocity in 10^6 m/s. Below, we tabulate several parameters for a MBH passing through a sun-like star. We use the density data from Solar interior given in [11]. Column 1 contains the fraction of Solar radius. Column 2 contains the gas density in 10^3 kg/m^3. Column 3 contains an estimated

R_{Sun}	ρ_3	v_6	F_t/M_{18_2}
0.0	146	1.39	$85 \cdot 10^9\ N$
0.1	82	1.33	$51 \cdot 10^9\ N$
0.2	35	1.19	$28 \cdot 10^9\ N$
0.3	12.3	1.06	$12.3 \cdot 10^9\ N$
0.4	4.0	0.96	$4.9 \cdot 10^9\ N$
0.5	1.35	0.87	$2.0 \cdot 10^9\ N$
0.6	0.49	0.80	$0.86 \cdot 10^9\ N$
0.7	0.185	0.74	$0.38 \cdot 10^9\ N$
0.8	0.077	0.69	$0.18 \cdot 10^9\ N$

Table 1.
Parameters for MBH passing through a sun-like star.

speed of a MBH arriving from a distance of thousands of solar radii. Column 4 contains F_t for $M_{18} = 1$ (**Table 1**).

The Solar radius is $R_\odot = 6.96 \cdot 10^8\ m$. Thus, we estimate the energy loss of a MBH passing through the center of a Sun-like star:

$$\Delta E = \int_{-R_\odot}^{R_\odot} F_t\ dx = \left(2.0 \cdot 10^{19}\ J\right) M_{18_2}. \tag{B.2}$$

Author details

Mikhail V. Shubov
Department of Mathematics, University of Massachusetts Lowell, Massachusetts, USA

*Address all correspondence to: mvs5763@yahoo.com

References

[1] Khriplovich IB, Pomeransky AA, Produit N, Ruban GY. Can one detect passage of small black hole through the earth? Physical Review D. 2008;77(6):1-6

[2] Pania P, Loebb A. Tidal capture of a primordial black hole by a neutron star: Implications for constraints on dark matter. Journal of Cosmology and Astroparticle Physics. 2014;**06**(2014):026

[3] Carr B, Kuhnel F, Sandstad M. Primordial black holes as dark matter. Physical Review D. 2016;**94**:1-32

[4] Bramante J. Dark matter ignition of type Ia supernovae. Physical Review Letters. 2015;**115**(14):1-6

[5] McDermott SD, Yu HB, Zurek KM. Constraints on scalar asymmetric dark matter from black hole formation in neutron stars. Physical Review D. 2012; **85**(2):1-24

[6] Abramowicz M, Becker J. No observational constraints from hypothetical collisions of hypothetical dark halo primordial black holes with galactic objects. Astrophysical Journal. 2009;**705**(1):659-669

[7] Beckmann RS, Devriendt J, Slyz A. Bondi or not Bondi: The impact of resolution on accretion and drag force modelling for supermassive black holes. Monthly Notices of the Royal Astronomical Society. 2018;**478**(1):995-1016

[8] Sánchez-Salcedo FJ, Brandenburg A. Dynamical friction of bodies orbiting in a gaseous sphere. Monthly Notices of the Royal Astronomical Society. 2001; **322**(1):67-78

[9] Ostriker EC. Dynamical friction in a gaseous medium. The Astrophysical Journal. 1998;**513**(1):252-258

[10] Bodenheimer PH. Principles of Star Formation. Berlin, Heidelberg: Springer Verlag; 2011

[11] Guenther DB, Demarque P, Kim Y-C, Pinsonneault MH. Standard solar model. The Astrophysical Journal. 1992; **387**:372-393

[12] Williams D. Molecular Physics. New York: Academic Press; 1961

[13] Chakrabarti SK. Accretion Processes on a Black Hole. Bombay, India: Tata Institute of Fundamental Research; 1996

[14] Chattopadhyay I, Sarkar S. General relativistic two-temperature accretion solutions for spherical flows around black holes. International Journal of Modern Physics. 2018;**28**(02):1950037

[15] Colpi M, Maraschi L, Treves A. Two-temperature model of spherical accretion onto a black hole. The Astrophysical Journal. 1984;**280**:319-327

[16] Kocsis B, Loeb A. Menus for feeding black holes. Space Science Reviews. 2014;**183**(1–4):163-187

[17] Tinyakov P, Kouvaris C. Growth of black holes in the interior of rotating neutron stars. Physical Review D. 2013; **90**(4):043512

[18] Contopoulos I, Gabuzda D, Kylafis N. The Formation and Disruption of Black Hole Jets. Berlin/Heidelberg: Springer; 2015. pp. 1-9

[19] Zhang B. The Physics of Gamma-Ray Bursts. Cambridge: Cambridge University Press; 2018

[20] Lide DR. CRC Handbook of Chemistry and Physics. 84th ed. Boca Raton, Florida: CRC Press; 2003

[21] Ambastha A. Probing the solar interior: Hearing the heartbeats of the sun. Resonance. 1998;**3**(3):18-31

[22] Hubbell JH. Photon Cross Sections, Attenuation Coefficients, and Energy Absorption Coefficients from 10 Key to 100 GeV. Washington D.C.: Center for Radiation Research National Bureau of Standards; 1969

Chapter 10

A New Hypothesis of Spin Supercurrent as Plausible Mechanism of Biological Nonlocal Interaction, Synchronicity, Quantum Communication

Yunita Umniyati, Victor Christianto and Florentin Smarandache

Abstract

We start with citing a seminal paper by Josephson-Pallikari-Viras, that biological entities can be assumed to be able to communicate nonlocally, i.e., instantaneously. However, they also admit that the underlying mechanism of such an entangled communication is not clear yet from the wave mechanical equations. Similar arguments have been pointed out by several authors, citing that quantum equations themselves have not described anything on a possible mechanism of quantum-type interaction between two biological entities. This chapter intends to fill that research gap by suggesting a new hypothesis of spin supercurrent as a physical mechanism, based on the assumption of macroquantum condensate having nonlocal effects. Moreover, we also draw several potential applications including superconductor quasi-crystalline structure of space and plausible new method of quantum communication. Such an argument is outlined herein partly based on our personal encounter with astrophysical quantization in the past 17 years or so.

Keywords: biological nonlocal interaction, quantum nonlocality, entanglement, spin supercurrent, superfluid dynamics, superconductor quasicrystalline, quantum communication

1. Introduction

In a seminal paper by Prof. Brian Josephson—Pallikari-Viras, they argued that despite quantum nonlocal interaction tends to be undetected by statistical averaging, but by assuming macroquantum system, biological entities can be assumed to be able to communicate nonlocally, i.e. instantaneously. However, they also admit that the underlying *mechanism* of such an entangled communication is not clear yet from the wave mechanical equations [1, 2].

IntechOpen

Actually, it is known for a long time that quantum physics allows quantum correlations—common reliance of attributes of wave capacity of supposed entangled quantum substances while there is space partition. This may include a phenomenon called the near-field antenna effect, i.e. the presence close to radio wire (a wavering electric dipole) of superluminally spreading electromagnetic field [3].

Nonetheless, various arguments have been pointed out by several authors, citing that quantum equations themselves have not described anything on the possible mechanism of quantum-type interaction between two biological entities.

In this chapter, we will discuss some existing literature and then we come up with a new hypothesis that spin supercurrent provides the sought-after physical mechanism for biological nonlocal interaction, synchronicity, and plausible new quantum communication method.

2. Literature survey

2.1 Wave mechanics equations

In quantum mechanics, the depiction of action of the field-free magnetic vector potential depends on Schrödinger's equations without presenting any actual interaction. As the activity of the field-free magnetic vector likely takes place in space where the electromagnetic field is missing, this potential has both non-electric and non-attractive nature. While there are researchers who did try to develop an electric representation of quantum wave mechanics, such as Gabriel Kron, but it did not give new results, as far as our knowledge. Moreover, in our previous book, Shpenkov and Kreidik have shown that Weyl provided cut-off to solutions of the original Schrodinger equation (3D), to achieve a quite good agreement with experimental data at the time. It is clear that in most textbooks on QM, whenever the authors discuss solutions of spherical Schrodinger equations, they rarely compare the results with actual experimental data, because they know, there is no agreement at all between spherical wave mechanics and experiment. It should be clear, that despite fairytale stories have been circulated to invoke certain mystical elements to wave mechanics origin, the fact is, it was a failed attempt since the beginning [4].

2.2 Classical EM theory approach

As Boldyreva wrote, which can be paraphrased as follows: "EM hypothesis portrays field-free magnetic vector potential. In traditional electrodynamics, the magnetic field of acceptance B is resolved by condition curl = B curl A , where A will be an attractive vector potential. In protecting of attractive field, 0 = B, the accompanying may happen: $0 \neq A$. This case is alluded to as the without field vector potential. Magnetic vector potential has its very own actual significance. In 1949, Erenberg furthermore Siday anticipated the capacity of attractive vector potential to impact straightforwardly the attributes of quantum substances despite the fact that there is no electromagnetic field at the area of the elements. In 1959, the chance of such an impact was considered by Aharonov and Bohm. Hence, an extraordinary number of tests have been directed which affirmed the hypothesis" [3].

2.3 Macroquantum condensate hypothesis

Here allow us to mention our chapter in a journal of Foundation Louis de Broglie 2006, suggesting that astrophysical quantization can be explained for instance by assuming macroquantum condensate of astrophysical bodies [5].

Provided such astrophysical macroquantum effects can be accepted, then it seems not so hard to suppose that under certain circumstances biological nonlocality interaction can happen, once we assume similar macroquantum condensates.

2.4 Spin supercurrent in superfluid helium

What we can read in some recent papers by Liudmila Boldyreva, she puts forward an argument of the existence of spin supercurrent to mediate biological quantum-type interactions [3].

Boldyreva wrote among other things, which can be paraphrased as follows:

> *"This work proposes basically another way to deal with portrayal of the above-thought about peculiarities: specifically, it is shown that it is conceivable to portray these peculiarities as far as such actual interaction as spin supercurrent. The twist supercurrent arises between objects having turn, and its activity will in general make equivalent the individual characteristics of precession of twists of collaborating objects. (Note that Yuri Bunkov, Vladimir Dmitriev and Igor Fomin were granted the Fritz London Memorial Prize in 2008 for the investigations of spin supercurrents in superfluid 3 He-B)."*

In this model, quantum correlations between quantum entities may be performed by spin supercurrent emerging between virtual photons (virtual particles pairs) created by those quantum entities.

2.5 Carl Jung's synchronicity

Limar wrote a review on the possible link between Carl Jung's concept of synchronicity with quantum non-local effect, known as entanglement. He argues in favor of cellular level or DNA level quantum type interaction, such as meiosis etc. Nonetheless, he also admits that many paper streams on this subject are plagued by the non-existence (as yet) of the physiological or physical mechanism of such non-local interaction [6].

2.6 Research gap

Similar arguments have been pointed out by several authors, citing that quantum equations themselves have not described anything on the possible mechanism of quantum-type interaction or communication between two biological entities. This chapter intends to fill that research gap by suggesting a new hypothesis of spin supercurrent as the physical mechanism, based on the assumption of macroquantum condensate having nonlocal effects. Moreover, we also draw several potential applications including superconductor quasi-crystalline structure of space and plausible new method of quantum communication. Such an argument is outlined herein partly based on our personal encounter with astrophysical quantization in the past 17 years or so.

3. A new hypothesis

If now we put all the above findings from macroquantum condensate (or close to superfluid 3 He) to spin supercurrent, hence, we come up with a new hypothesis, that we will state here for the first time:

3.1 Hypothesis

"There is spin supercurrent to be observed to mediate interaction between biological entities, between consciousness which known as synchronicity (in Jungian term), and also to provide quasi-crystalline structure of space, and in turn, it allows a new model of quantum-type nonlocal communication."

3.2 Simple physical model

According to Bunkov and Volovik, the superfluid current of twists—turn supercurrent—is another agent of superfluid flows, for example, the superfluid current of mass and molecules in superfluid 4He; superfluid current of electric charge in superconductors [7].

According to Boldyreva, such spin supercurrent mechanism can be helpful to mediate biological nonlocal interaction, can be modeled as follows:

$$J_z = g_1(a_1 - a_2) + g_2(t_1 - t_2), \tag{1}$$

where g_1 and g_2 are coefficients depending on deflection angles and the properties of the medium where spin supercurrent emerges. Turn supercurrent is certainly not an electric or attractive interaction and therefore it is not protected by electromagnetic screens [3].

Moreover, Boldyreva also argues that such a spin supercurrent interaction can find implications in alternative medicine, which can be paraphrased as follows: " a deterministic portrayal of the association is utilized, which continues as per the laws overseeing the conduct of sub-atomic fluid when the temperature of the last option is near outright zero (the properties of superfluid 3He-B). This methodology concurs with E. Schrödinger's perspective communicated in his book '*What is life?*', i.e. "The living organic entity is by all accounts a naturally visible framework which to a limited extent of its conduct ways to deal with that absolutely mechanical (as differentiated with thermodynamical) direct to which all frameworks tend, as the temperature moves toward outright zero and the atomic problem is taken out" [8].

4. Sideways and rationale

4.1 Our personal encounter with macroquantum condensate astrophysics

This segment permits us to recount an account of our experience with macroquantum condensate in astrophysics. Everything started by a to some degree "reasonable deduction" (or readers might call it: einfuhlung), when one of us (VC) got an old book by Nozieres and Pines [9], on superfluid Bose liquid. He inquired: Let us see what this book can bring to the domain of astronomy and cosmology. Before long, he tracked down many fascinating discoveries with regards to the writing, from W.H. Zurek to Grigory Volovik and so forth. That is the start of our undertaking for more than 18 years up to this point, coming about a few papers in a series [5, 10–14]. The soonest paper called "Cantorian superfluid vortex hypothesis" was distributed in January 2004, where VC presented a forecast of potential areas of three new circles of planetoids on the external side of Pluto. Then, at that point, after 2 years, VC distributed a paper in AFLB [5], where he laid out what are potential clarifications of macroquantum impacts in astronomy (for example, noticed likewise by Tifft and furthermore Virginia Trimble and so forth). One of the contentions in that AFLB paper is macroquantum condensate, for example, conceivable quantum impact actuated by BEC or superfluid-type medium [5].

More recently, we (VC, FS, YU) come up with an argument of cosmological entanglement supposing such a macroquantum effect is real.

4.2 Observational evidence

4.2.1 Quantization of planetary orbit distances in the solar system

In this section, we will review the work and results by us, during the past 17 years or so. The basic assumption here is that the Solar System's planetary orbits are quantized. But how do their orbits behave? Do they follow Titius-Bode's law? Our answer can be summarized as follows (**Figure 1**):

Navier-Stokes equations --> superfluid quantized vortices --> Bohr's quantization rule

Figure 1.
From NS turbulence to quantized vortices.

And it seems that the proposed model is slightly better compared to Nottale-Schumacher's gravitational Schrödinger model and also Titius-Bode's empirical law [1, 11].

The evidence of quantization of planetary orbit distances seem to suggest to wave mechanics model at a large scale [5, 10–14]. See also Peter Coles [15].

4.2.2 Observational finding on cosmological entanglement

Interestingly there is a recent report from MIT suggesting that ancient quasars support such quantum entanglement at large-scale phenomena. In an article, it is reported about the possibility of cosmological entanglement [16], which can be paraphrased as follows:

"In 2014, ... the William Herschel Telescope and the Telescopio Nazionale Galileo, both located on the equal mountain and separated via about a kilometer. One telescope focused on a particular quasar ..... Meanwhile, researchers at a station located between the two telescopes created pairs of entangled photons and beamed particles from each pair in contrary directions toward every telescope" [17].

Therefore, such a discovery has opened up a new way to look at the Universe: *an entangled Cosmos* [18, 19].

4.2.3 Newtonian action at a distance: Smarandache's

Hypothesis expresses that there is no speed limit of anything, including light and particles [20]. Eric Weisstein likewise composed ramifications of Smarandache's Hypothesis [21], which can be summarized as follows: "...the speed of light c is as of now not a biggest at which estimations can be sent and that abstract speeds of data or mass switch can occur. These confirmations fly notwithstanding every idea and investigation, as they misuse both Einstein's exceptional rule of relativity and causality and don't have any test support. It is genuine that current preliminaries have confirmed the presence of positive sorts of quantifiable superluminal quirks....." [21].

While the thought is very basic and in view of known speculation of quantum mechanics, called Einstein-Podolski-Rosen bridge, actually such a superluminal material science appears to be still difficult to acknowledge by the greater part of physicists. Beginning around 2011, there was a clear astounding outcome as declared by the OPERA group. Regardless, hardly any months after the fact, it was disavowed on the ground of mistakes in dealing with the estimation.

Permit us to offer not many remarks on such a clear inability to identify quicker than light speed as follows: Despite those discussions over the OPERA results, we believed that a seriously persuading test has been finished by Alain Aspect and so on; he had the option to show that quantum non-territory association is genuine. In 1980, Alain Aspect played out the first EPR try (Einstein-Podolski-Rosen) which demonstrated the presence of room nonlocality (Aspect 1982). Alain Aspect and his group at Orsay, Paris, led three Bell tests utilizing calcium course, i.e. the first and last utilized the CH74 disparity. The second was the first use of the CHSH imbalance.

The third (and generally well known) was organized with the end goal that the decision between the two settings on each side was made during the flight of the photons (as initially proposed by John Bell). A few experimenters demonstrated a comparative outcome until the distance of more than 90 km.

So, the thought of "spooky action at a distance" is a genuinely actual peculiarity. In addition, activity a way off was at that point referenced in Newton's *Principia Mathematica.* Regardless of obviously Einstein was attempting to make each of Newton's demeanors into nothing, our result suggests that the Maxwell equations in classical electrodynamics have "spooky interaction at a distance" type of interactions (as it has also been proven for Coulomb potential), which might be noticed both at limited scope tests just as in a cosmological scale, as ongoing confirmations show similar effect at a distance in relation to Smarandache's hypothesis.

4.2.4 Evidence of Cooper-pair tunneling in nuclei is likely to indicate superfluid vacuum model instead of gluon

In a recent report published in *Phys. Rev. C*, Potel et al. wrote on a breakthrough on the subject was made through the study of one- and two-neutron transfer reactions with heavy ion collisions in inverse and direct kinematics, enabled by the use of magnetic and γ-ray spectrometers, which suggest that there can be Cooper-pair tunneling in nuclei [22]. In retrospect, this finding seems to indicate that the superfluid vacuum model can be a better approach than the gluon model as in the Standard Model. See also [23, 24].

Besides, the superfluid nuclear matter hypothesis is known for a quite long time, especially going beyond BCS theory, cf. Walecka, Matsuzaki, Lombardo *etc.* [25, 26].

4.2.5 Initial evidence on the synchronicity between patient and doctor

In a 2008 article, Alex Hankey argues in favor of Macroscopic Quantum Coherence in Patient-Practitioner-Remedy Entanglement. An interesting remark in his article goes, which can be paraphrased as follows: "A different relationship length implies that the quantum cognizance's initially infinitesimal connection length currently becomes naturally visible. We reason that, for the most part, at every basic precariousness (remembering input dangers for natural administrative frameworks), quantum vacillation fields display plainly visible quantum cognizance" [27].

Although he did not come up yet with a clear physical mechanism of such macro-quantum coherence, one can arrive at a similar hypothesis of spin (supercurrent) interaction like Boldyreva's, as it is known in biological phenomena. See Likhtenshtein [28].

4.2.6 Initial evidence on galactic synchronicity

Although it is known that "One of the cornerstones of inflationary cosmology is that primordial density fluctuations have a quantum mechanical origin," as Kanno & Soda wrote, however, most physicists consider that such quantum mechanical effects disappear in CMB data due to decoherence [29].

We have discussed before that cosmological entanglement has been observed, which in turn, it can be attributed to the superfluid turbulent interstellar medium.

Presently, there is a new striking report by Charlotte Olsen et al., proposing that 36 cosmic systems appear to have "facilitated" in a such way that they seem to give synchronized stars arrangement. From Olsen et al. paper, they do not give a potential hypothetical explanation [30].

Notwithstanding, by theorizing such a twist supercurrent system likewise can occur at cosmic scale in view of superfluid interstellar medium, we can concoct a "potential" clarification, that such a lucid star arrangement is because of some sort of "*galactic synchronicity*." We know that such a term is not accessible yet in present cosmological vocabularies, however, we can predict that time for that term will come as well, as there is likewise a book, proposing that synchronicity is probably going to show up all around in Cosmos [31].

4.2.7 Other experimental results

Other reports seem to indicate that there are reasons to believe such a quantum effect between consciousness, mind-matter interaction, and also Aharonov-Bohm type interaction in the superfluid vortex [16, 32, 33]. Suter et al. also provide other experiment evidence, as they wrote in the abstract: "The data unambiguously show that Bzclearly deviates from an exponential law and represent the first direct, model independent proof for a nonlocal response in a superconductor" [34].

5. Discussion: four plausible applications in various fields

a. A new theoretical model of high-temperature superconductivity may lead to extremely efficient energy generation and transmission

b. A new type of electronic device.

c. Superconductor quasi-crystalline vacua hypothesis.

d. The plausible new method of quantum communication.

The explanation for each of the aforementioned plausible applications will be discussed shortly below:

5.1 A new theoretical model of high-temperature superconductivity may lead to extremely efficient energy generation and transmission

It is known that a superconductor permits the flow of current without resistance. The conventional way of thinking about the transition from normal to superconducting is called the Bardeen-Cooper-Schrieffer (BCS) theory. But last year, H. Koizumi, a researcher at Tsukuba University has announced a new theoretical model of high-temperature superconductivity, which may lead to extremely efficient energy generation and transmission. Instead of focusing on the pairing of charged particles, this new theory uses the mathematical tool called the Berry connection. This value computes a twisting of space where electrons travel. In the standard BCS theory, the origin of superconductivity is electron pairing. In this new theory, the supercurrent is identified as the dissipation less flow of the paired electrons [35]. We will discuss later; we may come up with an alternative method of quantum communication based on such Berry connection.

5.2 A new type of electronic device

Hua Chen et al. wrote experimental evidence which can lead to a new type of electronic device based on spin supercurrent, according to their abstract which can be paraphrased as follows: "In slight film ferromagnets with amazing simple plane anisotropy, the part of absolute twist perpendicular to the simple plane is a decent quantum number and the relating turn supercurrent can stream without scattering. In this Letter we clarify how turn supercurrents couple spatially remote turn blending vertical vehicle channels, in any event, when simple plane anisotropy is flawed, and examine the likelihood that this impact can be utilized to manufacture new kinds of electronic device" [36, 37].

5.3 Superconductor quasicrystalline vacua hypothesis

As we discussed in a forthcoming paper [38], we discuss on the possibility that the space consists of discrete cells, to become cells composed of superconductor quasi-crystalline. We put forth a new hypothesis that the discrete cellular structure of space consists of cells of superconductor quasi-crystalline. It is argued that the definition of quasicrystals should not include the requirement that they possess an axis of symmetry that is forbidden in periodic crystals. The term "quasicrystal" should simply be regarded as an abbreviation for "quasiperiodic crystal," possibly with two provisos.

To sum up, quasicrystals display a non-periodic, yet ordered, arrangement of atoms. They contain a small set of local environments which reappear again and again, albeit not in a periodic fashion. Their structure is not random either, since the diffraction pattern shows sharp Bragg peaks, although their symmetry is noncrystallographic, with the n-fold symmetries (n = 5, 8, 10, ...) stemming from the fact that these local environments occur with n equiprobable orientations. A recent discovery suggests that quasi-crystalline has a superconductive phase at a very low temperature [39].

It may resemble Finkelstein's hypercrystalline model of vacuum. What if the quasicrystalline model is not in semiconductor solid...but a superconductor quasicrystalline? We may call it: super-crystalline structure of 3D space.

Quasicrystalline solid is also good because it brings in more than three dimensions, which may be very relevant. This would also bring in Finkelstein and Penrose and some of Frank Tony Smith's investigations. The next item to consider is a super-quasi-crystalline solid (SQC).

Because of its fractal properties, we can expect that the SQC can extend down to the structure of space, similar to what Finkelstein envisaged [40].

The quasicrystal structure of space may be composed of solid matter or soft matter, of which its general dynamics have been outlined by Fan et al. [41]. Plausible assembly of soft matter quasicrystal is shown in **Figure 2**:

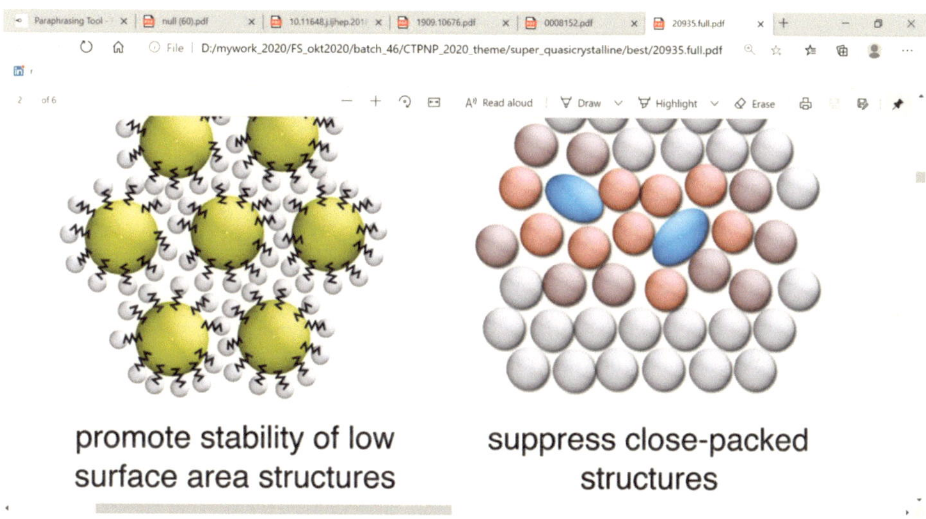

Figure 2.
Plausible assembly of soft-matter quasi-crystal.

It exhibits close-packed structures. This dense-packed structure of space should be verified with experiment. A few observables:

5.4 Natural quasicrystal in rock:

Steinhardt and Bindi [42] argued a unique hypothesis, proposing that quasicrystals might conceivably be pretty much as hearty and steady as gems, maybe, in any event, framing normally. These contemplations roused a very long term look for a characteristic quasicrystal finishing in the revelation of icosahedrite ($Al63Cu24Fe13$), an icosahedral quasicrystal found in a stone example made fundamentally out of khatyrkite (translucent $(Cu,Zn)Al2$) named as coming from the Koryak Mountains of far eastern Russia. In their paper, they contended that the examination demonstrates the example to be of an extraterrestrial beginning (**Figure 3**).

Figure 3.
Light image of entire MSNF specimen prior to sampling, Khatyrka CV3 meteorite [42].

Moreover, some papers argue that such a rock may be of manmade origin, as Bindi et al. noted:

> *"The proof for the presence of the quasicrystal deliberately work in the stone is thusly predominantly solid. Be that as it may, the perception of intermetallic compounds with copper and iron, which requires a profoundly lessening climate, is profoundly bewildering. It raises the likelihood that the example started from slag or another anthropogenic interaction. Nonetheless, the example was found in a far-off locale exceptionally a long way from any modern action." [43]*

While we admit that it would need further studies, as we see it such a hypothetical origin of meteorites and rock from extraterrestrial or manmade origin remains puzzling. It may be more possible to argue in favor that the quasicrystalline that happens in nature was caused by the structure of space itself is composed of SQC.

5.4.1 Natural quasicrystals in the solar system

Luca Bindi and also Matthias Meier et al. seem to suggest that quasicrystals have a cosmic origin [44, 45]. While such a hypothesis is quite reasonable, allow us to add a possibility that such a cosmic origin might yield from hidden structure of space itself. Such a hypothetical origin may be more "workable" than most quantum gravity hypotheses [46].

What is more interesting here is that Sakai has presented superconductor effects of quasicrystals [47].

5.4.2 A plausible new method of quantum communication

Inspired partly by Koizumi's research at Tsukuba Univ., we may come up with an alternative method of quantum communication based on such Berry connection. Prof. M.V. Berry is widely known for his research, but mostly for his theorem called Berry phase and Berry connection, which are often linked to the Aharonov-Bohm effect.

While there are various researchers who have come up with a number of possible quantum information or quantum internet methods, such as Pomorski & Staszewski, who wrote: "The idea of quantum web was.

displayed in this work. The more definite picture requires considering different impacts as decoherence processes that drive the quantum position-base qubit out of its cognizance just as decoherence processes that annihilate the cognizance of qEC (quantum Electromagnetic Cavity) [48].

Nonetheless, just yesterday an idea came to us, inspired by Berry connection and also crystalline structure of space. Actually, Tesla came up first with an idea to propagate telecommunication with the help of the telluric field of Earth, but alas it was canceled because Marconi obtained the patent first before him.

As we mentioned above, Sakai presented evidence of superconductivity of quasi-crystal, therefore provided, we accept the super-quasi-crystalline model of 3D space, we may come up with an idea that can be considered as quantum communication built on the crystalline structure of space itself. The outline of our idea is as follows, as we began to read papers by Prof. Michael V. Berry from UK (**Figure 4**):

Super-crystalline vacua -> Berry connection -> spin supercurrent -> nonlocal information transmission -> quantum communication system.

Figure 4.
A process flow of quantum communication based on Berry connection.

If we can prove this can work, at least a conceptual design, then may it can be a quite viable alternative to the 5G cellular network.

Moreover, such an assumption of superconductor crystalline structure of space, it seems to find support from our descriptive model of the Solar System in terms of superconductors. See our recent paper [49, 50].

6. Concluding remarks

In this chapter, we discuss how conventional wave mechanics does not provide a physical mechanism which is supposed to mediate nonlocal biological interaction, as discussed by Josephson-Pallikari Viras and others. Based on the hypothesis and also research findings by Yuri Bunkov, L. Boldyreva et al., we submit wholeheartedly that a new hypothesis of spin supercurrent is the sought-after physical mechanism, based on the assumption of macroquantum condensate having nonlocal effects.

In the last section, we discuss four plausible applications of such a scheme, including: a. A new theoretical model of high-temperature superconductivity may lead to extremely efficient energy generation and transmission; b. A new type of electronic device; c. Superconductor quasi-crystalline vacua hypothesis; d. Plausible new method of quantum communication. Clearly, more research is recommended to verify further what we outlined herein.

Acknowledgements

The authors would like to express sincere gratitude to Robert Neil Boyd, Ph.D., Prof. The Houw Liong, Ph.D., and numerous scholars for many discussions.
Version 1.0: 29th May 2021, pk. 15:01.
Version 1.1: 30th May 2021, pk. 16:25.
Version 1.2: 13th June 2021, pk. 14:30.
Version 1.3: 18th Jan. 2022, pk. 2:31.
YU, VC, FS.

Author details

Yunita Umniyati[1], Victor Christianto[2]* and Florentin Smarandache[3]

1 Department of Mechatronics, Swiss-German University, Tangerang, Indonesia

2 Satyabhakti Advanced School of Theology, Jakarta, Indonesia

3 Mathematics, Physical and Natural Sciences Division, University of New Mexico, Gallup, NM, USA

*Address all correspondence to: victorchristianto@gmail.com

References

[1] Josephson BD, Pallikari-Viras F. Biological utilization of quantum nonlocality. Foundations of Physics. 1991;**21**(2):197-207

[2] Josephson BD. The discovery of tunnelling supercurrents. Reviews of Modern Physics. 1973;**46**(2):251

[3] Boldyreva LB. Spin supercurrent in phenomena of quantum non-locality (quantum correlations, magnetic vector potential) and in near-field antenna effect. Journal of Modern Physics. 2019;**10**(2):128-144. DOI: 10.4236/jmp.2019.102010

[4] Sphenkov G, Kreidik L, Krasnoholovets V, Christianto V, Smarandache F. Old Problems and New Horizons in World Physics. New York: Nova Science Publishers; 2020. BISAC: SCI055000

[5] Christianto V. Annales de la Fondation Louis de Broglie. 2006;**31**(1)

[6] Limar IV. Carl G. Jung's synchronicity and quantum entanglement: schrödinger's cat 'wanders' between chromosomes. NeuroQuantology. 2011;**9**(2):313-321. ffhprints-00637383v2f

[7] Bunkov YE, Volovik GE. Spin Superfluidity and Magnon BEC. 2012. arXiv:1003.4889v3 [cond-mat.other]

[8] Koizumi H, Ishikawa A. Berry connection from many-body wave functions and superconductivity: Calculations by the particle number conserving Bogoliubov-de Gennes equations. 2021. arXiv:2105.02364v1 [cond-mat.supr-con]

[9] Nozieres P. Pines D. Theory of Quantum Liquids: Superfluid Bose Liquids. 1st ed. Boca Raton: CRC Press; 1994. p. 208. ISBN: 9780201408416

[10] Christianto V. Apeiron. 2004;**11**(1):112-152

[11] Christianto V, Rapoport DL, Smarandache F. Progress in Physics. 2007;**2**:9-11

[12] Christianto V, Smarandache F. Progress in Physics. 2008;**1**

[13] Christianto V, Smarandache F, Umniyati Y. Prespacetime Journal. 2020;**11**(3)

[14] Smarandache F. Christianto V. Progress in Physics. 2006;**2**:5-12

[15] Coles P. The wave mechanics of large-scale structure. In: The Early Universe and the Cosmic Microwave Background: Theory and Observations. Dordrecht: Springer; 2003. pp. 219-236. Arxiv: 0290576 (astro-ph)

[16] Kreskens C, Perez DL. Evidence of Quantum Consciousness in Evoked Zero-spin Echoes. bioRxiv preprint. DOI: 10.1101/219931

[17] Umniyati Y, Christianto V, Smarandache F. An explanation of Sedna orbit from condensed matter or superconductor model of the solar system: A new perspective of TNOs. AIP Conference Proceedings. 2021;**2331**(030014). DOI: 10.1063/5.0041656

[18] Friedmann A. Can the cosmos test the quantum entanglement? In: Astronomy. 2014. Available from: www.astronomy.com

[19] Genovese M. Cosmology and entanglement. Journal of Computational

and Theoretical Nanoscience. 2009;**2**(3):303-309. Available from: Researchgate.net

[20] Smarandache F. There is no speed barrier in the universe. Bulletin of Pure & Applied Sciences. 1998;**17D**(1):61 Available from: http://fs.unm.edu/NoSpLim.htm

[21] Weisstein EW. Smarandache hypothesis. The Encyclopedia of Physics. Available from: http://scienceworld.wolfram.com/physics/SmarandacheHypothesis.html

[22] Potel G, Barranco F, Vigezzi E, Broglia RA. Quantum entanglement in nuclear Cooper-pair tunneling with γ rays. Physical Review C. 2021;**103**:L021601

[23] Zhitnitsky A. Superfluid helium II as the QCD vacuum. Nuclear Physics B. 2017;**916**:510-549

[24] Zloshcastiev KG. Superfluid vacuum theory and deformed dispersion relations. International Journal of Modern Physics A: Particles and Fields; Gravitation; Cosmology; Nuclear Physics. 2020;**35**:2040032. DOI: 10.1142/S0217751X20400321

[25] Matsuzaki M. Dirac sea effects on superfluidity of nuclear matter. Physical Review C. 1998;**58**(6):3407. Available from: https://inspirehep.net/literature/4/5070

[26] Lombardo U et al. Superfluid states of nuclear matter beyond the BCS approximation. International Journal of Modern Physics. 2005;**14**(04):513-554. DOI: 10.1142/S0218301305003375

[27] Hankey A. Macroscopic quantum coherence in patient-practitioner-remedy entanglement: The quantized fluctuation field perspective. Evidence-based Complementary and Alternative Medicine. 2009;**6**(4):449-451. DOI: 10.1093/ecam/nen032

[28] Likhtenshtein G. Electron Spin Interactions in Chemistry and Biology. Switzerland: Springer International Publishing Switzerland; 2016

[29] Kanno S, Soda J. Bell inequality and its application to cosmology. Galaxies. 2017;**5**:99. DOI: 10.3390/galaxies5040099

[30] Olsen C et al. Star Formation Histories from SEDs and CMDs Agree: Evidence for Synchronized Star Formation in Local Volume Dwarf Galaxies over the Past 3 Gyr. arXiv:2104.06514v1 [astro-ph.GA]. 2021

[31] Durston S, Baggerman T. The Universe, Life and Everything. Amsterdam: Amsterdam University Press B.V; 2017

[32] Tremblay N. Independent re-analysis of alleged mind-matter interaction in double-slit experimental data. PLoS One. 2019;**14**(2):e0211511. DOI: 10.1371/journal.pone.0211511

[33] Polkinghorne R et al. Geometric Phases of a Vortex in a Superfluid. 2021. arXiv: 2101.07348

[34] Suter A et al. Direct observation of nonlocal effects in a superconductor. Physical Review Letters. 2004;**92**(8):087001

[35] Koizumi H. Superconductivity by berry connection from many-body wave functions: Revisit to Andreev—Saint–James reflection and Josephson effect. Journal of Superconductivity and Novel Magnetism. 2021;**34**(8):2017-2029. DOI: 10.1007/s10948-021-05905-y

[36] Boldyreva L. Alternative & Integrative Medicine. 2013;**2**:3. DOI: 10.4172/2327-5162.100011

[37] Chen H et al. Non-Local Transport Mediated by Spin-Supercurrents. 2014. arXiv: 1408.1100

[38] Christianto V, Smarandache F. Continuum hypothesis revisited: From cantor, to Godel, then to discrete cellular space model. Asia Mathematika Journal. 2022

[39] Kamiya K et al. Discovery of superconductivity in quasicrystal. Nature communications. 2018;**9**:154, DOI: 10.1038/s41467-017-02667-x. Available from: www.nature.com/naturecommunications

[40] Finkelstein DR et al. Hypercrystalline vacua. Available from: https://arxiv.org/abs/quant-ph/9608024

[41] Fan T-Y et al. Review on Generalized Dynamics of Soft-Matter Quasicrystals and Its Applications. arXiv: 1909.10676

[42] Steinhardt PJ, Bindi L. In search of natural quasicrystals. Reports on Progress in Physics. 2012;**75**(9):092601. DOI: 10.1088/0034-4885/75/9/092601

[43] MacPherson GJ et al. Khatyrka, a new CV 3 find from the Koryak Mountains, Eastern Russia. Meteoritics & Planetary Science. 2013;**48**(8): 1499-1514

[44] Bindi L et al. Icosahedrite, Al63Cu24Fe13, the first natural quasicrystal. American Mineralogist. 2011;**96**(5-6):928-931

[45] Matthias MMM et al. Cosmic History and a Candidate Parent Asteroid for the Quasicrystal-Bearing Meteorite Khatyrka. 2018. arXiv: 1803.03513

[46] Bindi L. Natural Quasicrystals: The Solar System's Hidden Secrets. Springer Nature: Switzerland; 2020 ISSN 2524-8596

[47] Sakai S. Superconductivity in quasicrystals. In: RIKEN Center for Emergent Matter Science. 2021

[48] Pomorski K, Staszewski RB. Towards quantum internet and non-local communication in position-based qubit. AIP Conference Proceedings. 2020;**2241**:020030. DOI: 10.1063/5.0011369

[49] Cramer JC, Herbert N. An Inquiry into the Possibility of Nonlocal Quantum Communication. 2014. arXiv: 1409.5098v2

[50] Umniyati Y, Christianto V, Smarandache F. An explanation of Sedna orbit from condensed matter or superconductor model of the solar system: A new perspective of TNOs. In: Paper Accepted at SMIC 2020 Conference, Held at Aug. 2020, Indonesia. AIP Conference Proceedings. 2021;**2331**:030014. DOI: 10.1063/5.0041656. Available from: https://aip.scitation.org/doi/abs/10.1063/5.0041656